T0330216

A JUST ENERGY TRANSITION

Getting Decarbonisation Right in a Time of Crisis

Ed Atkins

BRISTOL
UNIVERSITY
PRESS

First published in Great Britain in 2023 by

Bristol University Press
University of Bristol
1–9 Old Park Hill
Bristol
BS2 8BB
UK
t: +44 (0)117 374 6645
e: bup-info@bristol.ac.uk

Details of international sales and distribution partners are available at bristoluniversitypress.co.uk

British Library Cataloguing in Publication Data
A catalogue record for this book is available from the British Library

ISBN 978-1-5292-2095-7 hardcover
ISBN 978-1-5292-2096-4 paperback
ISBN 978-1-5292-2097-1 ePub
ISBN 978-1-5292-2098-8 ePdf

Cover design: Andrew Corbett
Front cover image: Getty Images/smartboy10
Bristol University Press use environmentally responsible print partners.
Printed and bound in Great Britain by CPI Group (UK) Ltd, Croydon, CR0 4YY

FSC
www.fsc.org
MIX
Paper | Supporting responsible forestry
FSC® C013604

For Jo

Contents

Acknowledgements

This book was written alongside numerous conversations with people working on, around, and against issues of decarbonisation, energy transitions, and social justice. I am grateful to the below for the generosity of their time and for their insight, expertise, and visions of what a just energy transition might (and should) be.

Ramón M. Balcázar, Observatorio Plurinacional de Salares Andinos

Albert Banal, Som Energia

Viktor Bjelic, Centar Za Zivotnu Sredinu

Liz Blackshaw, Trades Union Congress Northeast

Yvonne Blos, Friedrich-Ebert-Stiftung e.V.

Jeremy Brecher, Labor Network for Sustainability

Jack Breckenridge, European Marine Energy Centre

Helen Castle from the Rousay Eglisay & Wyre Development Trust

Ben Delman, Solar United Neighbours

David Deng, Voices for Power on Community Energy

Rebekah Diski, New Economics Foundation

Jennifer Donn, Rampion Offshore

Marianna Dudley

Ulrich Eichelmann, River Watch

Ian Epondulan, Voices for Power on Community Energy

Lewis Evans

Julian Felvinci, North American Megadam Resistance Alliance

Rebecca Ford

Denis Francisković, Eko Pan

Pippa Gallop, Bankwatch

Malene Lunden, Energi Akademiet

Zindzi Makinde, C3E International

J Mijin Cha

Jake Molloy, National Union of Rail, Maritime and Transport Workers

Ryan Morrison, Friends of the Earth Scotland

Sebastian Ordoñez Munoz, War on Want

Davina Ngei, Global Women's Network for the Energy Transition

Thuy Nguyen, Voices for Power on Community Energy

Caron Oag, European Marine Energy Centre

Debajit Palit, TERI India

Audra Parker, Save Our Sound

Martin Parker

Mark Pepper, Ambition Lawrence Weston

Christian Poirier, Amazon Watch

Aaron Priest, Viking Energy

Asha Ramzan, Voices for Power on Community Energy

Olimpiu Roib, Enercoop

Andy Rolfe, Schools Energy Coop

Chris Saltmarsh

Maria Sanchez-Lopez

Lara Santos, European Marine Energy Centre

Silvia Sartori, ENERGIA

Ewan Gibbs

Erik Hagen, Western Sahara Resource Watch

Hamza Hamouchene

Maureen Harris, International Rivers

Frank Hay, Sustainable Shetland

Annette Hollas, C3E International

Will Houghton, Bristol Energy Cooperative

Megan Howes, Interstate Renewable Energy Council (Solar Ready Vets)

Saskya Huggin, Low-carbon Hub

Gabrielle Jeliazkov, Platform

Harpreet Kaur Paul

Mat Lawrence, Common Wealth

Michal Levy, Voices for Power on Community Energy

Erin Savage, Appalachian Voices

Lara Skinner, The Worker Institute's Labor Leading on Climate Initiative

Jeremy Thorp, Ynni Teg

Adrian Warman, Thrive Renewables

Emily Watt and Anna Richardson, Bristol University Press

Patricia Miranda Wattimena, Asia Pacific Forum on Women, Law and Development

Andy Whitmore, London Mining Network

Joe Williams

Rebecca Windemer

Matt Wood, Energiesprong

Many of the chapters are in this book are accompanied by an illustration by Cai Burton, an artist based in Bristol in the southwest of England. This artwork stemmed from a process of collaboration and discussion with Cai. I would write a chapter, send it to Cai, and we would then meet to talk about its narrative. Cai's work and contribution to this project is present in more than the illustrations that come before each chapter. There were many moments when our discussions prompted a rethinking of how I approached this topic, the evidence I presented, and how I spoke about it. One such moment was when Cai presented preliminary concept sketches, including outlines of smokestacks and pollution. I was surprised by these. When Cai spoke about the chapter, his reading of my argument became clear. The focus of my draft had been solely on highlighting what is wrong with contemporary energy transitions and illuminating the injustices present within them. The language was critical and negative, and the mood of the chapter was dark. A positive vision of what a just energy transition might be was needed. As we sat with our coffees, talking to Cai signalled that my efforts in this work needed to focus on illuminating the successes of community-centred energy transitions, not just on what has gone wrong elsewhere. I am thankful to Cai for his artistic skill, energy, and engagement but also for calling me out at times like this. Conversations with Cai reminded me of what I wanted this book to be, and why it's important.

Introduction

A television host spins the wheel while another looks on. The two presenters of a popular daytime show in the UK have run this feature many times before. They call viewers and offer them the chance to win prizes through a lottery draw. Prizes usually include cash awards, split into portions of the wheel that, once spun, can be landed on. On 5 September 2022, there was a difference. These options, labelled onto glittering gold and bright-blue segments of the wheel, are '£1,000' and 'Energy Bills'. A reward at stake is the payment of £400 towards a viewer's energy bills, each month for four months. To select who plays, the presenters call viewers one by one, without warning and require those called to answer with a passphrase. Most days, this leads to a few false starts as viewers miss the call or fail to give the correct password. The first viewer called on this show has just got back from visiting their 93-year-old great aunt. He answers the phone with the set phrase immediately. The two presenters introduce the game and prizes, asking the caller: "How are your energy bills? Are you worried about it at all?" The viewer answers that they are reliant on a pre-payment meter, where users pay for energy before using it, and that they have major concerns about paying their bills into the winter. The wheel spins. It lands on 'Energy Bills'. The sound of the viewer's relief is heard down the phone line and across the studio.

This segment on national television represented the turning of many people's anxieties over spiralling energy bills into a gameshow format. Warmth is treated as a prize, like a new car or a holiday. Electricity as a reward to be won, rather than a right or human need. The competition was soon labelled as dystopian and tone-deaf in a context of a widening energy price crisis in the UK in 2022 (Salisbury, 2022). The game appeared on the daytime show for the following few days with 'Energy Bills' still offered as a prize. However, it was quietly removed within the week. When it returned to the show a month later, the prizes offered were cash awards only.

Competitions to win warmth and light are signals of the failure of how we currently manage energy. We often only become aware of just how much electricity we use, and how we use it, at moments when supply becomes precarious. Spiralling bills in 2022 were one such moment. In the UK, electricity bills increased by 54 per cent in April 2022 and a further 27 per cent in October of the same year (Bolton and Stewart, 2023). The latter rise was, to some extent, restricted by the Energy Price Guarantee enacted by the national government, which, in theory, limited the average household energy bill to £2,500.

We all have energy stories. These are linked to personal needs and wishes or the deliberations and debates within our households on how best to use electricity. Energy usage is a Sunday roast and spin-drying the laundry to ward off damp. It is kids laughing as they play their games or leaving the light on overnight for a child scared of the dark. In the face of rising bills, these energy stories changed. They became formed of slow cookers, air fryers, and microwaved soup. Fairy lights and candles. Home-made draught excluders from old pairs of jeans. Extra jumpers, thicker jumpers. Smart meters and new thermostats. Double, triple, or secondary glazing. Loft and cavity wall insulation. New boilers and electric heaters. These are just a few of the things that friends, family, colleagues, and strangers mentioned when I told them that I was writing this book. All signal the failure of our current energy model.

Rising prices have demonstrated the inability of the UK energy sector to provide warmth and light to many. This has, at times, been absurd. In January 2022, the UK energy company, Ovo Energy was forced to apologise after advising customers to cuddle pets and do star jumps to keep warm (BBC, 2022a). At other points, the stories have been tragic. In March the same year, UK media outlets reported the heart-breaking story of an elderly woman in Harrogate, Yorkshire, who was so anxious about rising energy bills that she travelled on a bus all day (Rodger, 2022). When asked about it on national television, the then Prime Minister, Boris Johnson (2019–2022) boasted that she was able to do so because of his government's policy of giving free bus passes to the elderly (Morris, 2022). Across the UK, many have been forced into skipping meals so that they can pay the gas bill, or take on second or third jobs just to keep the lights on.

Further price rises will have a disproportionate impact on those most vulnerable to such changes, such the elderly. Around 2.8 million older households (three in every ten) in the UK are estimated to have been pushed into energy poverty in 2022, with household income after paying energy bills falling below the national poverty line (Age UK, 2022). It will also impact those on lower incomes particularly harshly because they must dedicate a higher percentage of their spending to essentials than others, meaning that they are especially vulnerable to any price rises (ONS, 2022b). Between January and July 2022 alone, 187,000 customers, who had problems paying rising bills, were moved by their energy providers onto a pay-as-you-go meter, in which they will pay higher prices (Citizens Advice, 2022). Many, powerless against spiralling costs of living, turned to 'self-disconnecting' and going without topping up a pay-as-you-go electricity meter entirely. In 2022, an estimated 3.2 million people couldn't afford to top up the meter and, as a result, were cut off from their energy supply (Citizens Advice, 2023). This represents one person losing energy every ten seconds of the year – each of them a tragic illustration of the plight of many during a prolonged cost-of-living crisis. However, this is also a crucial reminder that access to energy is an essential need regardless of how much somebody might be able to pay for it.

There's no promise of a return to normality any time soon. Some 78 per cent of households in the UK use gas central heating to keep warm (BEIS, 2022a). This widespread reliance means that many of us are extremely vulnerable to changing energy prices and uncertainty driven by national and international events, such as the violent invasion of Ukraine by Russia in 2022 which heavily disrupted flows of gas from Russia into Europe. In the UK, the energy price cap was introduced in 2019 to limit what individuals paid on the standard variable tariffs that are offered as the default option by energy providers. It, in theory, caps the price a supplier can charge per kilowatt hour of electricity or gas used, with prices reviewed every six months. While it was never seen as permanent (intended to end in 2023), the energy price cap has been a lot more variable than might have been expected at first. In response to spiking wholesale prices, the cap was increased in February and August 2021, and again in April and October 2022. The price cap was replaced by an 'Energy Price Guarantee' in October 2022.

The energy model, as it stands, is not fit for purpose. These changes have occurred at a time of growing anger at the profits made by those providing energy. Energy companies continued to make profits during this crisis. Centrica's profits increased to £1.34 billion in the first half of 2022 (Giordano, 2022). It is estimated that energy generators and providers in the UK may make excess profits of £170 billion between 2022 and 2024 (Wickham and Gillespie, 2022). The National Grid's annual pre-tax profits increase by 107 per cent in the 2021–2022 financial year, equalling £3.4 billion (Phillips, 2022). All while the prices paid by energy users spiralled.

There has been organised action against this unfairness. Campaigns such as Enough is Enough, backed by national trade unions, have called for fair pay and taxation, lower energy bills, and better access to housing. The Don't Pay campaign group was organised by collecting individual and household pledges to cancel payments to energy companies on 1 October 2022, the date when bills were forecast to rise dramatically. This campaign echoed similar payment strikes against the introduction of a poll tax (a blanket tax applied to every eligible individual, regardless of income or wealth) in the late 1980s. It grew and gained steam throughout the second half of 2022. Polling showed that 1.7 million households intended to cancel their energy bills (Ellson, 2022). The campaign gained 200,000 direct pledges from bill-payers that they would refuse to make payments (Mathers, 2022). While this number didn't fulfil an initial ambition of one million pledges, the campaign exercised a pivotal influence on energy politics in the UK in 2022.

The Don't Pay campaign represents an expression of the dissatisfaction that many people hold with the energy sector in the UK. The mass pledge to cancel payments also shows a degree of leverage that customers can have in challenging the sector and in, ultimately, influencing government policy (Milburn, 2022). The campaign influenced the subsequent policies of the government led by Prime Minister Liz Truss (2022) to limit energy prices going into October 2022. In a presentation to ministers at the UK government Department for Business, Energy and Industrial Strategy (BEIS) in August 2022, bosses from E.ON presented the 'Don't Pay' campaign as an existential threat that would have led to a further failure of the sector (Williams, 2022). According to the company, one million households refusing to pay their energy bills would have led to a collective loss of £265 million a month for those companies providing energy in the UK (Williams, 2022). Notably, E.ON reported a pre-tax profit of over £3 billion for the first half of 2022 (Patrick, 2022).

Within this context of crisis, a transition to renewable energy is no longer just about addressing or slowing climate change. It is also about the cost of living and national energy security. Whereas environmentalists once pushed for renewable energy investments because it was *greener* than burning fossil fuels, they now show that it is both *cheaper* and more *secure* too. Political narratives of energy security, particularly in the countries of the European Union and the UK, have triggered an expansion of renewable energy generation and energy efficiency measures. A 2022 energy strategy announced in Germany accelerated the decarbonisation of its energy grids, pledging that 80 per cent of total electricity consumption would come from renewable power by 2030. This was done using a new language in global energy transitions, asserting that the expansion of renewables was in the public interest by ensuring future energy security.

All of this takes place against a backdrop of climate breakdown impacting places and communities across the globe (IPCC, 2022a). As I have written this book, vast floods have damaged and destroyed communities in India, Bangladesh, Brazil, Pakistan, and South Africa. Wildfires have ravaged swathes of Afghanistan, France, Siberia, and Portugal. Hurricanes and cyclones have hit Mozambique, Madagascar, and the Philippines. Animals are experiencing mass die-offs. Other species have disappeared into extinction. Heatwaves affected much of Western Europe. On 19 July 2022, the mercury may have gone as high at 40.3°C in Coningsby, Lincolnshire. In Spain, temperatures reached 45.7°C.

Addressing climate breakdown necessitates the decarbonisation of our energy systems. This is understood as the move away from the use of fossil fuel energy sources that emit the greenhouse gases that are causing climate change. The Intergovernmental Panel on Climate Change has called for the rapid deployment of renewable energy technologies (such as wind and solar) and for new policies to encourage behavioural shifts in how we use and manage electricity (IPCC, 2022b). This requires the transformation of how we generate and use energy for electricity and heating or cooling. There is urgency to climate action. Emissions reductions must happen comprehensively and must happen now. Yet, current policy interventions often focus on the short-term aims of reducing greenhouse gas emissions – and using renewable energy transitions as a tool to do so. The focus is often on 'transitioning' away from fossil fuels to low-carbon sources of energy, not on what might make such a change 'just' or 'fair'. It is no longer a question of *if* renewables will replace fossil fuels. Renewables are expanding and here to stay. It's a question of *how* we conduct this change.

A just energy transition

Renewable energy enjoyed a year of record growth in 2022, with an estimated 320 gigawatts (GW) of new generation, including solar, wind, hydro, and bioenergy power, added globally (IEA, 2022a). Such figures would have been near-unthinkable a decade ago, with those calling for such a scale of transition likely being dismissed as utopian or, worse, delusional. Solar power is now the cheapest electricity source in history, with lower costs of the technology leading to more and more projects being built. Fossil fuels may be displaced by a 100 per cent renewable energy grid within the next 25 years (Way et al, 2021). In May 2022, 50 per cent of electricity in the UK came from sources deemed 'low-carbon' (National Grid, 2022). Coal power provided less than 0.1 per cent to the generation mix. High winds on 24 May 2022 helped drive this, generating 19.9 GW of electricity – a national record (National Grid, 2022). Offshore wind and solar energy are expanding in the UK: in 2022, the government's flagship renewables scheme approved future renewables schemes that would lead to enough electricity to power almost 12 million homes (close to 11 GW of electricity) (BEIS, 2022b). This was close to double the capacity approved in previous rounds of the scheme. These projects will be followed by many more. If wind farms were built on all available and appropriate land in the UK, they could provide for an estimated 140 per cent of the national energy demand (Enevoldsen, 2019). If the UK was to pursue renewables at such a scale, it would find itself generating a national energy surplus and exporting electricity to other countries (Gosden, 2022).

Energy transitions do not happen on a blank canvas. While decisions related to the decarbonisation of energy are made at the international, national, and regional levels, they take form locally. Across the globe, many communities live at the sharp end of decarbonisation. This includes those living in areas and regions where renewable energy infrastructure is being built. It also encompasses those who you might not necessarily think of when we first speak of renewable energy transitions. These are households facing energy insecurity and poverty, living in cold homes but unable to afford the measures that can address this. Or the residents of towns and regions that are reliant upon carbon-heavy work for employment. Or the communities living in the 'lithium triangle' of Bolivia, Chile, and Argentina, who face the impacts of the lithium mining that underpins renewable technologies. While there is much talk of fossil fuel assets and investments being left cut off and stranded by decarbonisation, there is also a need to protect potentially 'stranded' and 'excluded' communities across the globe and provide them with new futures.

This book is organised around the call for a *just energy transition* – the argument that, for any energy transition to be successful, it must give people a new future that is better than the present. Decarbonisation is often presented in purely technical terms, discussed in a language of gigawatts and emissions savings. However, it is political and social. It is characterised by investment flows, decision-making, and the distribution of benefits and impacts. Renewable energy projects are subject to a process of consultation and opposition in which local and national priorities, concerns, targets, and stories coalesce. It is through these processes that decarbonisation creates new energy landscapes of national targets and expansive energy grids, lost jobs and new employment, built infrastructure and changing terrains, altered local histories and experiences, and the creation of new winners and losers.

What can often be missed in energy transition policies is that the low-carbon alternatives need to be as good – and preferably better – than what they set out to replace. A just energy transition doesn't just bring about decarbonisation. It also involves altering and restructuring the energy system itself, sharing benefits and recognising the impacts and costs of any transition. In times of crisis, policy choices and options can often focus on low-hanging fruit, such as keeping the lights on or trying to keep energy bills low. Yet, with energy (and energy prices) touching all parts of life, such 'easy' solutions can overlook the transformative potential of an energy transition. When talking about energy transitions, we need to illuminate, understand, and speak about how they offer an opportunity to both avert climate change and redesign an energy system that includes, supports, and empowers people and their communities. An energy transition must be done *with* and *for* people, households, and communities, rather than something that happens *to* them.

While climate campaigners are seeking to draw attention to the 'end of the world', many people are worried about the cost of living today and how they might make it to the end of the month. This contrast has provided ample material for those wishing to slow energy transitions or deny their importance. Decarbonisation in the UK has been presented by critics as undemocratic and unfair, and as having disproportionate costs on poorer households who remain unprotected against spiralling energy prices (Atkins, 2022). As Lee Anderson, an MP from the Conservative Party who represents the constituency of Ashfield in Nottinghamshire has put it:

> Mention the Net Zero journey to people in Ashfield and they will look at you as if you have just arrived from another planet ... Many of them are only just managing to make ends meet. They want to be able to switch on the heating without worrying about the next bill. (Anderson, 2022)

There are flaws to this argument. Anderson alludes to Net Zero being a driver behind increasing bills. Yet, rising energy bills in 2022 were mostly the result of a combination of higher wholesale prices caused by cold winters increasing demand, historically low levels of gas storage, and precarious global supply because of the Russian invasion of Ukraine. Broader decarbonisation policies and changes implemented earlier may have reduced energy prices: the slowing of decarbonisation by Prime Minister David Cameron from 2013 onwards increased household energy bills by up to £2.5 billion (Evans, 2022a).

Despite this, Anderson's words are important. He is making an argument for energy justice, highlighting an apparent disconnect between ambitions of decarbonisation and climate action with the lived realities of many people whose lives will be changed. These narratives draw from anxieties in households and communities about rising costs of living. They also highlight important complexities of decarbonisation, linked to how current energy transitions fail to take the lived realities, hopes, or anxieties of people affected into account. Gary Smith, the general secretary of the GMB, a general trade union with over 460,000 members, argued in 2022 that those calling for the expansion of renewables represent a 'bourgeois environmentalism' that has 'no interest in jobs for working-class communities' (Wearmouth, 2022). This argument would be addressed by a more-expansive focus on the social good that energy transitions can bring, rather than a sustained characterisation of decarbonisation to reduce greenhouse gas emissions only.

In this book, I also make an argument of justice that asserts that any energy transition must not merely focus on the decline of fossil fuels but should also be dedicated to making our energy grid better, more equitable and more inclusive. This builds on the work of others, such as Dayaneni (2009), who have argued that a narrow focus on decarbonisation as emissions reduction risks a 'carbon fundamentalism' in which issues of equality and exclusion remain overlooked and unaddressed. Energy transition policies should pursue both decarbonisation and ensuring that any energy grid works for everybody, rather than prioritising the former over the latter. Broadening decarbonisation to address entrenched inequalities would empower broader social movements, allay the anxieties of those at the sharp end of decarbonisation, and address the complaints of those, who, like Anderson and Smith, claim to speak for them.

All transitions are messy and difficult, representing a move from one set of circumstances to another and creating new winners and losers. The decarbonisation of the energy grid can be grossly unfair, perpetuating dynamics that exclude some communities, impact others, fail to recognise different approaches or motivations, and overlook entrenched inequalities (Bouzarovski, 2022). Energy supply and demand link together vulnerable communities at the energy periphery, workers at power stations, and individuals at the light switch. At every stage of this 'energy continuum', a form of injustice can be found (Hernández, 2015; Newell, 2021).

The following chapters explore how renewable energy transitions, as they are currently formed, need to make more space for people and communities. In arguing for a just energy transition, I draw on previous work on the topics, concepts, and vocabularies of 'just transition' and 'energy justice'. A language of *just transition* originated in trade unions' call for better workers' rights and protections. It has gained considerable weight in climate politics, policy, and research to help illuminate how decarbonisation can be about more than just reducing greenhouse gas emissions, it can also be about protecting jobs and regional economies. The terminology of *energy justice* provides a multi-dimensional understanding of how access to energy is often skewed and monopolised along the lines of social, economic, and political power. Both terms provide an important challenge to claims that energy transitions are technical pursuits only. I adopt them to recentre energy policy and decision-making on a focus on equity and equality, in which policy seeks to redress past harms, include communities in decision-making, and support people in any process of decarbonisation.

This book presents a vision of what a just energy transition should be. This involves both national and regional governments and local communities taking a more-expansive role in any future decarbonisation. A just energy transition must be formed of energy policies and schemes that put people first and do so through both top-down government led approaches and bottom-up, community-centred policies. While the former can be located within current government plans to reduce greenhouse gas emissions, the latter involves people getting together in their communities and defining their own vision of an energy transition. The coexistence of the two would allow the large-scale energy infrastructure needed to ensure future energy security to sit alongside local, place-based schemes that allow people to take an active role in decarbonisation, rather than be treated as passive recipients of renewable energy.

If the future is renewables, it must be brighter. Such a future requires a political and economic transition as much as a technical one. Renewable energy is a common resource. We share it. Unlike fossil fuels, it is freely available across places and communities. However, the infrastructure that catches this energy and uses it to power our electricity grids is increasingly privately owned. This infrastructure encloses renewable energy and sells it for profit. If shareholder owned companies remain powerful in energy transitions, the benefits of decarbonisation are more likely to merely line the pockets of their shareholders, rather than help stimulate further positive transformations (Stewart, 2022a).

The UK energy system, as it stands, is not capable of decarbonisation while ensuring equity and justice. This is due to the dominant logic that the sector's structure represents: in which many utilities are shareholder owned and guided by profit. Privately owned companies hold an expansive influence over today's energy transitions. The UK is one of the only countries in Europe to have fully privatised its transmission grid. The other country where the sector is almost entirely privatised is Portugal – where energy bills have remained cheaper than in the UK.[1] The National Grid, responsible for transmitting all electricity in the UK, is a private company, and one of the largest investor-owned utility companies in the world.

Energy in the UK is primarily supplied to households by privately owned companies including British Gas (owned by Centrica), E.ON UK (owned by the German-based, investor-owned E.ON SE), Ovo Energy (primarily owned by its CEO, Stephen Fitzpatrick alongside other institutional investors), and Scottish Power (a subsidiary of the Spanish multinational Iberdrola, significant shareholders of which include the Qatar Investment Authority). Other energy companies are owned by foreign governments – EDF Energy UK, a leading company in national energy supply, is owned by Électricité de France which is, in turn, 85 per cent owned by the French national government. One exception is Octopus Energy, owned by the Octopus Group which is a private business albeit 75% owned by its employees (Octopus Group, 2023).

The energy price crisis has exposed the failings of the model of a private investor-dominated energy sector. In previous years, the potential for profit led many new entrants into the energy market, often with limited expertise or knowledge of the sector itself. An example of this can be found in the story of Avro Energy, set up in 2016 by 24-year-old Jake Brown, previously an amateur footballer, using a family loan. The offer of cheap energy bills to customers drove the company's growth, with it gaining over half a million customers between its founding and 2021. The company was primarily funded through customers paying their bills in advance, with the company's owners (Jake Brown and, his father, Philip) seeming to invest little of their own money. Avro Energy also directed payments of £4.25 million (described as 'management charges') to companies owned by the two Browns, as well as lending them £700,000, and an additional interest-free loan of £830,000 to a property company that they owned (Gosden and Brown, 2021; Davies, 2022).

Between 2018 and 2021, Citizens Advice (2021) raised concerns with Ofgem, the UK energy regulator, about Avro Energy – including unease about billing errors, how debts were being recovered, and the blocking of customers who were looking to move to other energy companies. Ofgem did little to address these concerns or use its enforcement powers to change Avro's business practices (BEIS Committee, 2022). When the company was established, there was also an absence of any check of whether the Browns were 'fit and proper' people to be leading an energy company. Rising gas prices exposed the flaws in the company's business model, including the absence of a strategy to manage any risk to supply or increasing prices. Avro Energy was soon unable to pay wholesale prices with the money that they raised from advance payments made by customers. The company made substantial losses and collapsed in 2021, with the company owing £90 million to its customers, who were later transferred to a different provider (Davies, 2022).

The story of Avro Energy was used by MPs in the UK to demonstrate the failings of the energy sector. The company was one of 28 energy companies that collapsed in 2021. This number included Bulb which was, at the time, the UK's seventh largest energy supplier, serving 1.7 million customers. Bulb was founded in 2013, receiving funding from venture capital firms who saw the company as having a high potential for growth and future profits. The company advertised itself as providing all its energy from renewable sources – with this claim, savvy marketing, and incentives to recommend the company to friends and family, driving the company's rapid growth. In 2021, Bulb was ranked by the *Financial Times* as one of the fastest growing companies in Europe (Kelly, 2021). The company went bust later that year.

The UK government was forced to intervene, with Bulb deemed 'too big to fail' and placed in special administration and bailed out. Rather than customers moving to other energy suppliers, this process ensured that there was no change to supply, with the government overseeing a process of restructuring. The cost of saving Bulb has been estimated at £2.2 billion (Pickard and Thomas, 2022). Its sale in mid-2022 only attracted one bidder: the investor-owned, 'Octopus Energy'. The British government bailed out a failed energy company, only to sell it back to private investors.

The current logic of the UK energy sector – of growth, profits, dividends, and private ownership – is incompatible with hopes for a just energy transition. As Matt Huber (2022) has argued, this is for two primary reasons. First, decarbonisation requires central, government-led, and public planning that prioritises the roll-out of renewable energy infrastructure *across* the grid. Second, successful energy transitions demand expansive spending regardless of the associated costs or financial risk. Private energy providers' primary motivation (and loyalty) is to their shareholders, who they serve by maximising profits and driving down costs. In seeking to increase profits, private capital is also restless, seeking larger returns through new markets, technologies, and infrastructures. While many fossil fuel companies have announced plans to 'go green' and invest in renewables, profitability can often remain the pivotal variable in defining these interests' role in an energy transition. This desire for profitability can often lead many energy providers, investors, and decision makers to neglect demand-side measures, like energy efficiency, that could make people's lives better and their electricity bills cheaper but do not chime with the desire for profit (Newell, 2021).

Fossil fuel energy companies have also in the past worked to stymie decarbonisation and, at times, promote climate denial, due to how action might affect their profit margins (Stokes, 2020). In the US, energy utilities spent an estimated US$554 million (£450 million)[2] on lobbying against national climate policy between 2000 and 2016 (Brulle, 2018). The spending of such sums dwarfs the capacity of those calling for renewable energy and, as a result, restricts the possibilities of energy transitions. There is a similar experience in the UK, where energy companies have lobbied the government to ensure that gas extraction remains part of any energy transition (Kennedy, 2021). Centrica, the owner of British Gas, has been linked to the funding of climate denial groups in the US, and lobbied for and funded key fracking interests in the UK (Carter, 2016; Hope, 2017; Hall, 2018).

The model of private ownership of energy contrasts with high levels of support for the public ownership of energy in the UK. Some 59 per cent of those polled by YouGov in December 2021 voiced support for national ownership. Such a move won't necessarily solve the of energy insecurity. Nor will it directly lead to decarbonisation. However, it will create a new context for decarbonisation. *We Own It*, a campaign calling for public ownership of energy and other services in the UK, argues that such a move would lead to an estimated saving of £3.7 billion a year from household bills in the UK (weownit.org.uk, nd.). It has been estimated that a move to public ownership of energy in the UK might reduce bills by 25 per cent – and that's on costs *before* the price rises of 2021–2022 (Hall, 2019). If private ownership of the energy sector came to an end, the money paid in energy bills can be returned to the sector, stimulating further investments in improving access and affordability.

Public ownership of the energy sector can often be presented as absolute, with a wholesale change of the grid. This can, in turn, lead to it being characterised as government overreach or misguided intervention. In 2019, the Labour Party in the UK presented a vision of publicly owned energy networks in its election manifesto. These plans were criticised as impractical and likely leading to declining services (Moore, 2019). Labour's nationalisation plans, which also included public ownership of internet infrastructure, were questioned in election coverage. This was often to the point of absurdity. In one moment, Angela Rayner, the then Shadow Education Secretary, was asked by an election debate host if the party also planned to 'nationalise sausages' during a discussion of climate change and meat consumption (Clarke, 2019). In 2021, the new leader of the Labour Party, Sir Keir Starmer, asserted that the party was committed to not nationalising the energy sector (Buchan, 2021).

This political context has been changed by rising energy prices in 2021 and 2022. At its 2022 party conference in Liverpool, the Labour Party announced Great British Energy. This proposed public-owned energy company would not be tasked with supplying electricity. Instead, its primary function is to exist as an investment vehicle to fund new renewable energy projects and work with energy companies to support the growth of wind and solar power in the UK. This was part of a broader suite of policies put forward by Labour to stimulate decarbonisation, including a target to decarbonise electricity supply by 2030 and an £8 billion national wealth fund to invest in low-carbon industries (Bell, 2022; George, 2022).

In the US, the 2022 Inflation Reduction Act of President Joe Biden has dedicated US$391 billion (£318 billion) to energy security and climate action, offering extensive subsidies and rebates to make low-carbon energy cheaper and reduce energy bills. This follows increasingly vocal calls for a Green New Deal, a policy approach that contains the explicit recognition that climate action and social justice are interlinked. The Green New Deal has become a popular slogan and demand of environmental movements – gaining support from social movements and campaigning groups, including Greenpeace, Extinction Rebellion, Friends of the Earth, and the Sierra Club. It has formed the basis of many recent books calling for policy action to not only mitigate climate change but to also address economic challenges, such as inequality (Aronoff et al, 2019; Klein, 2019; Pettifor, 2019; Ajl, 2021). A call for a Green New Deal asserts that any energy transition from fossil fuels necessitates policies and investments that put justice and equity first – and that the national government is best placed to lead such a process.

Both the UK Labour Party's plan for Great British Energy and President Biden's Inflation Reduction Act represents, to differing extents, a bringing of the state and national government back into energy transitions. These forms of a public-led energy transition don't necessarily equal a full return to government-run, centralised energy. It can, instead, take different forms, such as the collective, community ownership of energy infrastructure that allows for local approaches and participation (Cumbers, 2016). The European Union's Renewable Energy Directive (RED) (approved in 2009, updated in 2018, and revised in 2021) has provided an important framework to enable energy plans that are owned by local members, allowing for an expansion of community energy (Hoicka et al, 2021). Under this scheme, renewable energy community infrastructure must be controlled by members who are near the project itself but would be supported by legislation from national governments (Leonhardt et al, 2022).

Most visions of a new energy model are grounded at the local level and speak in terms of decentralised, local energy generation led by community organisations or energy cooperatives. Community and locally grounded energy generation schemes are not new. Current forms have roots in efforts in the 1970s to develop alternative technologies and a 'soft energy path' of energy efficiency and renewable sources (Lovins, 1976). In the wake of blackouts in New York City, Amory Lovins (1977) argued that centralised energy networks (with power generation sited in only a few locations and electricity transmitted over great distances) made the sector vulnerable to disruption. A new resilience of the network could be found in smaller, decentralised systems that provided electricity to local or regional areas.

'Community' energy can be broadly understood as local-led efforts to move beyond the current model of energy generation and supply by creating new routes to generate, develop and own energy infrastructure. In 2021, there were 123 community energy organisations active in the UK, generating electricity, providing energy efficiency support and advice, and supporting building improvements (Community Energy England, 2022). The sector has a rich potential in the UK and has grown rapidly over the past decade. In 2021, 217,400 people were involved across community energy organisations and £21.5 million was raised in investment for the sector (Community Energy England, 2022). The same year, the Environmental Audit Committee (2021), a select committee in the House of Commons which scrutinises the environmental credentials of government policies, argued that: 'Successful community energy projects across the country are placing tackling climate change at the heart of their activities, engaging the public in decarbonisation and pioneering innovative trials to meet the demands of a potentially decentralised and increasingly digitised future energy network' (Environmental Audit Committee, 2021). It has been estimated that the UK national government providing an investment of £1 million of technical assistance to each local authority in the country could create a cycle of investment that might lead to a total of £15 billion being invested in local energy networks (Sugar and Webb, 2022). Community-centred energy projects have myriad benefits. They can reinvest profits into local benefit funds, create new jobs and boost local communities, and work to address energy poverty. Localised energy generation might also be seen as more resilient than the centralised infrastructure of fossil fuels, with different networks able to autonomously create their own energy away from insecure global supply. Many people think local leaders are best placed to define climate action (LGA, 2021). However, these local authorities often lack the funding and capacity to commit to meaningful action (Barrass, 2022). Nor do these authorities necessarily have decision-making power and authority over the sectors that they might seek to decarbonise, like energy or transport.

However, the relationship between community energy and energy justice is not as clear-cut as it might seem at first (van Veelen, 2018). Community energy organisations aren't always totally democratic at the local level – they might prioritise different motivations and aspirations, creating suspicion around how resources are used and distributed, or privilege certain forms of expertise over democratic process (Simcock, 2016; Emelianoff and Wernert, 2019; van Bommel and Höffken, 2021). While local-led, community-based energy initiatives might provide answers to some of the various injustices discussed in this book, they may not be available and accessible to everyone.

An energy transition cannot spring from the local level on its own. An overarching focus on localised infrastructure does not necessarily match the scale of action required to decarbonise energy grids in the face of climate breakdown. We don't have the time available to decarbonise national energy grids one community at a time (Huber, 2022). Expansive transmission systems and battery storage technologies require huge levels of investment, which can only really take place in a highly centralised way. National action is necessary. Government-led policies must be implemented to direct, support, and scale up local energy transitions. As Hermann Scheer (2007: 232–233), widely seen as an architect of decarbonisation policies in Germany (broadly labelled 'Energiewende'), has argued, energy transitions can seek to create new geographies of energy supply and demand: 'By linking energy production and usage locally or regionally, it is possible to avoid the complex technical, organizational, administrative and political (including military) costs that both nuclear and fossil energy make unavoidable as they take their lengthy trip from production to final consumption' (Hermann Scheer, 2007). National policy in Germany empowered local energy networks, with 'Energiewende' policies including a process of market redesign in which new groups and organisations were incentivised to generate electricity (Pegels and Lütkenhorst, 2014). This opened up energy generation to smaller, local organisations, as well as municipally owned companies. The importance of these policies can be seen in how the number of new community energy producers slowed markedly when state subsidies were later withdrawn. Similar support from the national government in the UK is no longer forthcoming. The success of community energy has been slowed by regulatory barriers and a lack of government support. Community energy is mentioned only once in the 2020 Energy White Paper, published by the government of Prime Minister Boris Johnson, with this reference being a case study of a successful scheme in Bethesda, North Wales.

Decarbonising our energy grids is an opportunity to recentre the energy grid around communities. A just energy transition should not just be on reducing the emissions of the energy grid. It is a call that any transition should empower people and communities, ensuring that they can engage in the process and that benefits are shared equally.

A starting point

A call for a just energy transition represents an opportunity to do better, to improve lives, and to move beyond an energy system where warmth is a prize to be won on national television. The starting point of this book is to put people and communities back in the frame of decarbonisation. In a time of an energy crisis, it is necessary to put people's voices, stories, and experiences at the forefront of future decarbonisation policies. Each chapter begins with a journey to a location in the UK where decarbonisation is taking form and where a particular vision of just energy transition might be found. These trips have taken place with the COVID-19 pandemic in the background. My initial hope was to travel further afield, but successive waves and lockdowns hindered such possibilities. However, I have found great inspiration in these sites. Some are close to home: the community wind turbine in Lawrence Weston, Bristol (Chapter 5), is walking distance from where I live. Others have a personal connection: the Rampion Wind Farm off the coast of Sussex (Chapter 4) is within sight of beaches I often visited as a child, and the Cleve Hill Solar Park (Chapter 2) is found between towns where I spent time as a student in Canterbury. Some stories detail injustice, others describe emergent alternatives. All provided an opportunity for me to connect with decarbonisation taking material form, and to see and experience the landscapes, waterscapes and cityscapes that are being changed. It is these connections that guided the illustrations by Cai Burton that accompany each chapter, which present images of people holding up a frame, with an image inside of what decarbonisation *might* be – and how it can be better that the current energy model.

These sites and journeys went some way to illuminate different justices and injustices of contemporary energy transitions. The following chapters are organised around these different forms of justice and how a new energy model of both state-led policy and local, community-centred energy schemes might go some way to addressing them. Chapter 2 outlines this conceptual starting point by drawing from work on just transition, energy justice, and energy democracy to introduce the importance of a just energy transition. It focuses on expanding these frameworks to assert the importance of a nationally led but community-centred approach to any future energy model.

Chapter 3 focuses on *distributive injustice*, in which the scale of the benefits of renewable energy are enjoyed in one place but the social and environmental impacts are outsourced to different communities and other landscapes. The scale of action required creates a mismatch, with local-led schemes being too small to make a dent in the action required and bigger schemes overlooking issues of justice. Any just energy transition must include policies that allow people and communities to participate meaningfully. In the light of this, Chapter 4 explores *procedural injustice*, arguing that contemporary wind energy projects fail to benefit local communities living nearby and can exclude them from decision-making. It argues that the wind is, ultimately, public and that wind turbines should be put to use for the benefit of those living nearby.

Chapter 5 uses *justice as recognition* as its starting point to highlight how communities and local institutions can play various roles in decarbonisation. Renewables can become embedded in our everyday lives and economies. Community and local energy projects can challenge the current status quo of energy transitions and perhaps provide a better, more inclusive form of decarbonisation that is linked to local wealth generation.

The final three chapters highlight particular geographies of a just energy transition. Chapter 6 presents decarbonisation at the household level, highlighting how energy transitions can, if done right, represent an opportunity for restorative justice. With energy poverty and energy efficiency measures representing an overlapping challenge in an era of high energy prices, policy needs to prioritise the use of retrofitting and rooftop solar policies to support the most vulnerable. Solar panels and energy efficiency work can help people, reduce bills, and address energy poverty. A just energy transition can give people more than just light and heat. It can give comfort and dignity.

Chapter 7 explores how any energy transition is underpinned by work. Green jobs are priorities in national and international policy but there is limited understanding of what forms these jobs might take. This chapter explores what green jobs currently look like, who has access to them, and how they might develop in the future. Decarbonisation is a dramatic process of economic restructuring. Sectors will be transformed, work practices altered, and patterns of employment shifted. It is necessary to ensure that the workers who are forced to transition from fossil fuel work to low-carbon jobs are supported in such a move and that their new work is good, as well as green. For this to happen, it is necessary to recognise that workers have a place and role in defining an energy transition.

Chapter 8 widens the focus further, highlighting the important global dimensions of any energy transition. Renewable energy infrastructure is one point in a broader web of supply chains and logistics. Injustices occurred at every point of this web. Adopting a lens of cosmopolitan justice, this final chapter highlights how decarbonisation *here* joins communities and people with others *elsewhere*. Supply chains and commodity flows cross borders, tying together water pollution in Bolivia, the international transmission of electricity across the Mediterranean Sea and the disposal of electronic waste in Ghana. All such sites of a global energy transition highlight how, in short, there can be no transition *here* without justice *everywhere*.

If done right, a new energy model can create a just energy transition by helping vulnerable households and empowering communities across the globe. Over the coming chapters, I detail the different ways this might happen and present key rules for such a model of energy transitions. Taken together, these rules signal an approach that would allow for an energy transition that puts people first and recentres decarbonisation around a new, better energy model. These six rules call for all future energy transition policies to:

1. Push for community-scale energy projects.
2. Elevate and emphasise the participation and voices of communities.
3. Foreground community energy schemes in local economies and wealth building.
4. Prioritise those most vulnerable to energy poverty.
5. Ensure the participation and inclusion of workers.
6. Recognise that a just energy transition *here* must equal justice everywhere.

Transition

Starlings form their murmuration across Graveney Marshes at dusk, where mudflat meets greyish blue sky. They gather on telegraph polls and rooftops until a critical mass is found. Walking along this stretch of the Kent coastline between Faversham and Whitstable, you can see the tidal channel of the Swale that separates the Isle of Sheppey from the mainland. Beyond that, lies the Thames Estuary and the Kentish Flats Array, a collection of 30 offshore wind turbines, each 115 metres tall. The electricity produced by other offshore wind turbines along this coastline, such as at the London Array further in the distance, makes landfall on these marshes, where it joins overhead lines and the national grid. Crossing underneath these lines, you can hear the crackle from the electricity running elsewhere.

The nearby town of Seasalter is mentioned in the Domesday Book, as owned by the kitchens of Canterbury Cathedral and home to 48 smallholders, two 'ploughlands',[1] eight fisheries, areas of woodland and one church.[2] Today, the landscape is populated by apple trees and polytunnels where strawberries and raspberries grow. In the distance, a spritsail barge works its way across the channel. Walkers, day-tripping from London, move past herds of sheep and the flash of yellow and green of greenfinches flitting through a young crop of wheat. Birdwatchers seek out great white egrets, marsh harriers, peregrines, or warblers. All stop to watch over the marshes. Nearby lies the site of the last ground battle on British soil – a skirmish fought between the crew of a downed German plane and a group of soldiers billeted at the nearby Sportsman pub in September 1940.

'Project Fortress', formerly known as the Cleve Hill Solar Park, will soon cover this land. The project, due to be the biggest in the UK, will be formed of 19 banks of solar panels that will generate enough electricity to power more than 91,000 homes and the battery storage needed to stockpile it (Cleve Hill Solar, nd).[3] Due to its capacity exceeding 50 megawatts (MW) of energy generation, the Cleve Hill project is classified as a Nationally Significant Infrastructure Project. This means that final approval was given by the national government in February 2020 – rather than local authorities, as would usually be the case.

On my walk in this landscape in August 2021, Graveney, the closest village to the scheme, was dotted with posters calling for *'No Solar Power Station'* and *'No Battery Storage'*. The opposition to Cleve Hill incorporated numerous elements of dissent, ranging from ecological concerns to aesthetic disruption caused by the glint and glare of sunlight landing on the panels (Roddis, 2020). Concerns raised in the planning process included biodiversity impacts, the disruption caused by traffic and construction, safety, and the sheer size of the project.[4] The developers of this project have asserted its economic and ecological benefits, with a reported £27.25 million of investment providing jobs and opportunities for residents (Cleve Hill Solar, nd). Yet, opposition to the scheme centred on how the community was connected to the place and landscape itself and how such intricate links were not given space or value within the planning process. The importance of this landscape was often linked by those opposing the scheme to biodiversity and ecological value; to the birds who nest across the marshland and how personal perceptions and experiences of these marshes are intricately tied to the landscape, its avian visitors, and its broader biodiversity.

Many opponents to Project Fortress were clear that they did not wish to challenge the expansion and use of renewable energy, such as solar. Instead, they saw the project's 'green' credentials legitimising significant local impacts in the name of national energy transitions (Coward, 2018; Hutchinson, 2021). While impacts are local, the benefits of low-carbon energy are transported elsewhere by the overhead power lines. This leads to a sense of unfairness and injustice, with local communities unable to influence decisions but facing impacts of an energy transition but not enjoying any sense of benefit.

New energy landscapes

As governments adopt more targets to decarbonise energy supply, new 'energy landscapes' will emerge that embody the changing relationships between residents, communities, infrastructure, energy networks and beyond (Bridge et al, 2018; Bridge and Gailing, 2020). The infrastructure required by an energy transition must be built somewhere. While national action from state governments will lead any energy transition, it will have tangible impacts in certain areas. Changes will be physical but also highly social, political, and economic. They will be dynamic and complex, with local connections to landscapes, work, and community interacting with national and international targets to decrease emissions (Power et al, 2016). It is from these connections that dissent, such as that at Graveney Marshes, springs.

It is at schemes like 'Project Fortress' on the Kentish coast that communities are forced to confront the tension between supporting renewable energy and experiencing the impacts that it may have on their lives and landscapes. Despite popular support for renewable energy investments being high, many projects are subject to challenge and delay. The areas affected by new energy projects are often rural, with communities on the peripheries often facing changes and impacts to provide electricity to cities a great distance away (Naumann and Rudolph, 2020). Within these regions, opposition to and the acceptance of renewable energy projects is intricately linked to a sense of connection to a place and an appreciation of local biodiversity, community, and the landscape itself. It is also linked to a certain model of managing energy transitions that prioritises the interests of some over others.

Energy and energy networks are simultaneously personal and political. Energy touches every element of our lives. It powers our homes and workplaces, supports our movement around towns and cities, and underpins our digital lives and cultural goings-on. It is rare that we do something in our lives without a plug being switched on somewhere. Energy networks are also political in how they define the relationships between workers, communities, landscapes, institutions, and energy. The emergence of new ways of energy generation have always been inscribed with a particular way of structuring its generation, transmission, and management that enables the rewriting of relations of power and authority in society. The rise of coal as an energy source during the industrial revolution was linked to the desires of rich industrialists to concentrate production and workers in certain locations (Malm, 2016). In the UK, the mining of the material has held an expansive, social, economic, and cultural influence over lives and livelihoods – as has the extraction of crude oil in the North Sea (Beynon and Hudson, 2021; Marriott and Macalister, 2021). Internationally, energy retains an important geopolitical role. The historic influence and colonial power of the UK and later the US have been based on control of the dominant energy resources of coal and oil respectively (Ediger and Bowlus, 2019). Today we see the emergence of super-rich 'petrostates', such as Qatar, the United Arab Emirates (UAE) and Venezuela which are economically dependent on the extraction of fossil fuels. Some, such as Qatar and the UAE, have used their carbon-stained revenues to gain new cultural power, buying football clubs and hosting global tournaments. Others, such as Venezuela, have experienced severe boom and bust cycles leading to economic and political turmoil. Russia and some others have used their fossil fuel power as a blunt geopolitical tool, tightening the faucet to lessen the flow of fuel and influence others.

The energy landscapes to be formed by decarbonisation are as political as they are technical. While the new energy technologies that support decarbonisation may be presented as scientific, they are designed and built, implemented, directed, managed, maintained, and opposed or accepted by people operating social, economic, political, and cultural contexts. This can be seen in how renewable energy technologies can hold geopolitical importance and become embroiled in international trade wars. The rapid rise of solar manufacturing in China, driven by state subsidies, led to retaliation from both the United States and the European Union – who applied heavy tariffs on solar panels imported from China (Chaffin, 2012; Cardwell, 2014). The political character of decarbonisation is also evident in how energy policies are devised and enacted within particular social structures and are, as a result, underpinned by certain ways of understanding energy problems and solutions (Delina, 2018; Galvin, 2020). This context allows powerful interests and groups to dictate the shape of energy transitions, and who benefits (Sovacool et al, 2019a). These include:

Regulatory elites of policy makers, regulators, and politicians who use legal and political means to push for certain policies or to restrict others. Energy policies have provided fertile ground for the powerful to demonstrate or re-establish the authority of governments, centralising energy infrastructure and access or seizing land from local communities to secure resources and the revenues they can provide (Sovacool et al, 2019a; Dunlap and Arce, 2021). Across the globe, renewable energy projects have involved the displacement of communities and the seizure of their lands, often with limited routes for appeal (Dunlap, 2021a). This does not include the construction of renewable energy infrastructure alone, but also transmission lines and pylons, the growing of crops to be used as biofuels, and the extraction of ores and rare metals used in electric batteries.

Physical elites of law enforcement, criminal networks, and military forces who, either formally or informally, deploy violence and physical force to influence change or appropriate energy resources. Environmental defenders opposed to large hydroelectric projects have been subject to violence and, in some cases, murdered. These include Saw O Moo, shot dead on his way home by the Burmese military; Carlos Hernandez, murdered in his office in Tela de Atlántida, Honduras; Ricardo Pugong Mayumi, shot by assassins in the Philippines; and Jomo Nyanguti, shot by police officers at a dam site in Kenya (International Rivers, nd.). All were murdered in 2018 alone. In Oaxaca, Mexico, the construction of wind farms has coincided with a rapid increase in violence from both militarised police (in both an official and extra-judicial fashion) and violent *Sicarios* (hitmen, whose crimes in the region include kidnapping and up to 25 murders between 2009 and 2020) (Dunlap and Arce, 2021). Such violence occurs in a space where renewable energy infrastructure meshes with struggles for the autonomy of Indigenous communities, violence between organised crime groups, contests between workers and employers, and the financial motivations of all of these (Dunlap and Arce, 2021; Ramirez and Böhm, 2021).[5]

Technical elites of researchers, engineers, architects, and scientists can hold certain ways of understanding 'progress' or 'success' and put such ideals to work in defining what problems exist and what solutions are available. These groups often present decarbonisation and renewable energies within a narrative that is focused on technical aspects – of how to produce, store, transmit, and consume 'green' and 'clean' energy. A commitment to such a definition can result in a narrow understanding of both problems and solutions that come under the umbrella of decarbonisation. This might fail to allow space for or consider the wide variety of variables and impacts of energy transitions and how they both exist alongside and exacerbate historic and entrenched patterns of exclusion, inequality, and injustice and create new ones (Cederlöf, 2020; Sovacool et al, 2021). For example, understanding issues of energy poverty as exclusively a 'technical' problem of insecure supply being outpaced by demand neglects the deep-set social inequalities that define and limit individual energy use and access (Walker and Day, 2012).

Finally, *financial elites* of investors or business interests and owners might use the economic power of financial investment to influence which policies are supported or not. With workers dependent upon energy infrastructure for their livelihoods, the interests that own and operate these sites have huge power and can hold jobs ransom when faced with new regulations (Semuels, 2017). Fossil fuel companies have historically obscured the facts about climate change and slow policy making (Conway and Oreskes, 2012; Bonneuil et al, 2021). Those opposed to decarbonisation continue to assert a role for fossil fuels and even discuss climate action within broader culture wars, arguing that decarbonisation would cost too much or put jobs at risk.

Financial elites also have an important role in defining the forms that decarbonisation might take – with current policies and plans framed in terms of the green economy, in which pollution can be monetised, carbon emissions marketised and the private sector provides the investment and solutions required (Fletcher et al, 2019). For many financial organisations and interests, renewable energy infrastructures have come to represent safe and long-term revenue streams (Baker, 2021). The provision of state subsidies also encourages action and investment. This may not always be a good thing. In the United States, tax subsidies have, in essence, caused investment in renewables to become a form of tax shelter for some corporation's broader income – leading to these entities retaining an outweighed influence in national energy transitions. Key investors in renewable energy in the US include JPMorgan Chase, Google, Amazon, and the Bank of America (Knuth, 2021).

Financial motivations have also encouraged the entrance of organised crime groups into energy transitions. In Sicily, extensive subsidies for renewable energies led to the infiltration of the sector by the Mafia that was so expansive that roughly one-third of the island's wind farms were seized by authorities in 2013 (Bump, 2013). That year, some €1.3 billion was also seized from Vito Nicastri, a man accused of being the frontman of an operation that involved 43 wind and solar companies and 98 properties (BBC, 2013). Nicastri was later dubbed by the Italian media as 'Signore del Vento' (*Lord of the Wind*) (La Sicilia, 2019). In 2021, wildfires across Sicily were also linked to the Mafia, who were believed to have set fires to clear land to be used for solar PV farms (D'Ippolito, 2021).

In all of these cases, contemporary energy transitions risk being defined and dictated by the already powerful and wealthy and for their own political or financial benefit. Yet, decarbonisation can become more than just reducing greenhouse gas emissions. It can also represent an opportunity for society to become more socially inclusive – to empower communities, to include those who have previously been neglected and excluded. For this to happen, more must be done to understand the injustices and inequities in energy grids and how they might be addressed and transformed. To do so, this book adopts the conceptual vocabulary and ambition provided by writers on a Just Transition, energy justice and energy democracy – approaches that I define in the next section.

Broadening a just transition

The roots of the concept and language of 'just transition' are found in attempts by trade unions to reconcile ambitions for workers' rights and job protections with concerns for the environment and climate. As a workers' demand it was first articulated by uranium miners in Canada in the 1960s, but it was popularised by the work of Tony Mazzocchi of the US Oil, Chemical and Atomic Workers International Union (OCAW) in a plea for a 'Superfund' of government spending to protect workers (Galgóczi, 2020).[6] Mazzocchi and OCAW are celebrated as driving forces behind the 1970 Occupational Safety and Health Act in the US, which regulated that workplaces should be free from health hazards . This came in the face of toxic working environments, environmental pollution, and the closure of plants and factories, but it also provided a route to build and sustain coalitions between environmental groups and organised labour (Just Transition Research Collaborative, 2018). Mazocchi and other trade unionists recognised that their work was environmentally damaging but also affirmed that addressing these impacts did not have to mean the loss of jobs and upheaval for communities. The call for a just transition represents environmentalism that builds coalitions between climate activists and workers, by challenging the perception of environmental protection as a 'jobs killer'. As Brian Kohler of the Communications, Energy and Paper Workers Union, an early adopter of the concept asserted in 1996: 'the real choice is not jobs or environment. It is both or neither' (Kohler, 1996).

In the face of climate breakdown, calls for a just transition assert the need for climate action to make space for workers in any future. The just transition framework has stimulated a relative boom in academic work. Research has explored just transitions in the coal, automotive and transportation sectors (Abraham, 2017; Snell, 2018; Schwanen, 2020; Pichler et al, 2021). Others have detailed the vulnerability and experience of communities who are left 'stranded' by any transition away from fossil fuels, such as in the Ruhr Valley, Germany, New South Wales and Victoria, Australia, and Appalachia, US (Evans, 2007; Galgöczi, 2014; Snell, 2018; Snyder, 2018). Work elsewhere has detailed how this framework is (and might be) implemented in policy (Pollin and Callaci, 2019; Mayer, 2018; Snell, 2018; Snyder, 2018; Jenkins, 2019). Others have charted how it can be linked to (and learn from) broader concepts and movements, such as those for environmental and climate justice or fossil fuel divestment (Farrell, 2012; Evans and Phelan, 2016; Healy and Barry, 2017; Routledge, 2018).

A language of just transition is evident across the vocabulary of global climate politics. It has entered key documents of international organisations, such as the International Labor Organisation (ILO), the United Nations Environmental Programme (UNEP) and the United Nations Framework Convention on Climate Change (UNFCCC). The 2015 Paris Agreement affirmed the 'imperatives of a just transition of the workforce and the creation of decent work and quality jobs in accordance with nationally defined development priorities' (UNFCCC, 2016: 2).[7] At, the global climate meeting, COP24 in Katowice, Poland in 2018, the *Silesia Declaration* asserted the necessity of understanding – and addressing – the links between climate action and the rights and futures of workers. This declaration represented an important moment for the concept of just transition, with countries demonstrating a commitment to including workers in climate policy and action and the concept asserted as more than merely an add-on to policies that have already been made (Jenkins, 2019). Four years later, in May 2022, the Labour and Employment Ministers of the G7 countries signed a joint declaration that committed the countries to a shared just transition, ensuring that decarbonisation would include new, good and green jobs, social protections for those affected, and better workplace safety (ILO, 2022). Both statements show that a just transition is no longer a niche call by activists. It has become an ever-more central part of global climate politics and plans for a more sustainable future. It has become a broader policy priority, in which governments seek to devise and implement policies that lessen the impacts of climate action on workers and communities.

Despite this rise, the definition and application of a just transition remain broad and variable (Morena et al, 2020; Abram et al, 2022). The International Labor Organization (2015), the central arena for labour discussions on the international stage, issued the following guidelines for what a just transition should include:

- national macroeconomic and economic growth policies that prioritise employment;
- the use of environmental regulations in targeted sectors and industries;
- the creation of an enabling business environment for emergent 'green' enterprises;
- the presence of social protection policies to ensure the resilience of workers against the negative impacts of climate change, and mitigation and adaptation measures;
- policies that pursue job creation, limit job losses, and manage transitions to greener policies;
- occupational health and safety policies;
- skills development;
- social dialogue that ensures that all involved and affected by change and policies have a voice; and
- policy coherence, including the wide presence and coordination of sustainable development principles and social dialogue across different policies.

The translation of a just transition framework into practice is dependent upon government action and intervention (ILO, 2015; Healy and Barry, 2017; Cha, 2020). There are historical precedents for this, with many moments where governments have introduced comprehensive policies to support workers and communities facing upheaval. These have often taken the form of regional commissions that seek to support workers facing the decline of the industry that employs them. An often cited example of just transition is how the regional government of North-Rhine Westphalia in Germany's Ruhr region managed to both protect employment and support workers leaving declining carbon-heavy industries (such as coal miners and iron and steel workers) by implementing bridging payments for those close to retirement (providing wages for up to five years to support workers waiting to claim their pensions) and the provision of new training and personal development programmes to help younger workers find new jobs (Galgósczi, 2014). Subsequent schemes in the following decades sought to diversify the region's economy by attracting high-tech businesses and, in recent years, manufacturers in wind and solar energy supply chains (Galgósczi, 2014). Similar regional approaches to support workers transitioning away from carbon-heavy work include the Appalachian Regional Commission, formed in 1965, to support workforce and community development through funding key projects, proposed by locals facing the decline of the region's coal industry (Snyder, 2018).[8]

Policies that support workers have also been implemented by national governments and institutions. In the US, the Trade Adjustment Assistance programme, first implemented in 1962, provided financial support to workers who lose work, hours, or wages due to increased imports caused by trade globalisation. The programme includes components linked to skills and training, relocation allowances and support in job searches (Apollo Alliance and Cornell Global Labor Institute, 2009). More recently, in Spain, a Just Transition Strategy was launched in 2019 to protect communities in its historic coal regions from problems caused by the disappearance of fossil fuel work. This strategy was a key pillar in the government's integrated National Energy and Climate Plan – which focused on the social costs and benefits of decarbonization as well as just the environmental or climate ones (Martínez Rodríguez, 2021). It has been accompanied by close dialogue and collaboration between trade union groups and the private owners of the coal pits themselves to provide financial support for communities – including environmental restoration work, early retirement for miners over the age of 48, and the re-training of younger workers who lose jobs (ETUC, 2018; WRI, 2021).

These just transition policies of worker protections and support have an important parallel with recent UK policies adopted during the COVID-19 pandemic. In the face of COVID-19 causing a need for a national lockdown, the then Chancellor of the Exchequer Rishi Sunak announced the Coronavirus Job Retention Scheme on 20 March 2020. This policy implemented a furlough scheme, in which the government provided grants to employers, who were forced to close, to pay 80 per cent of staff wages and employment costs, up to a total of £2,500 each month. Other schemes were formed in 2020 to support workers whose working hours were reduced and those who were self-employed. The furlough scheme was initially anticipated to be active for three months. However, as the pandemic continued, it was ultimately extended to 30 September 2021. 11.7 million jobs were furloughed at 1.3 million businesses at a cost of £70 billion (Clark, 2021; ONS, 2021). One in four workers were furloughed during this period – with younger people, single working parents, those with disabilities, and those with lower levels of formal education more likely to be furloughed (ONS, 2021). Sectors particularly affected included hotels and restaurants, entertainment and the arts, and retail (ONS, 2021). In providing such extensive support, the 2020 Coronavirus Job Retention Scheme demonstrates the potential for national policy intervention to support businesses facing financial losses due to lost revenues, ensure secure incomes for those staff who were unable to work, and avoid mass redundancies and bankruptcies. It is an important example of how government intervention can ensure the resilience of individuals and businesses in the face of uncertainty.

Policies that are explicitly labelled just transition approaches have continued to emerge. In 2018, the Scottish government created a dedicated

Just Transition Commission, tasked with advising the devolved executive on how a low-carbon economy could be inclusive, ensure fair work and tackle inequalities. The commission met with workers and others involved in multiple lines of work (including the energy, transport, and construction sectors) – tracing the different ways in which the sectors may be impacted by climate action, and how climate policies can provide new opportunities. It issued its report in March 2021, summarising its work in four suggestions: a managed transition that creates opportunities and benefits for people across Scotland; a focus on skills and education; the empowerment of communities and local economies; and the broader sharing of benefits and burdens, based on the ability to bear such costs (SJTC, 2021). At the time of writing, the recommendations of the Commission had been delegated to different teams working in the Scottish government for evaluation and implementation.[9] In July 2022, a second Scottish Just Transition Commission (2022) released a new report, which called for more clarity from the Scottish devolved government on its decisions and highlighted the need for broader industrial policy to raise investment in large scale decarbonisation, to not only protect jobs but to improve them, and to tackle inequalities in energy transitions.

All of these policies illustrate the importance of national government in addressing the negative impacts of any transition away from fossil fuels. Central government institutions have the jurisdiction, financial resources, and political influence to enact the broad plans needed to protect and transition workers who are affected by decarbonisation. Just transition policies will often vary depending on national governance and institutional structures, the need for intervention and the perspectives of policy makers, activists, and communities (Barca, 2015; Goddard and Farrelly, 2018; Stevis, 2018a). This variance can be seen in the typology of policies defined by the Just Transition Research Collaborative (2018), with interventions (be they from regional or national institutions) understood as representing one of the following:

Status Quo: Policies that address climate change and provide support for workers but adopt a policy toolbox that preserves the current structure and rules. This might include the creation of new markets or the provision of incentives to businesses or consumers to stimulate new flows of investment and behaviour change. Or it might encompass the provision of corporate training, education, and relocation programmes. Governing institutions and processes remain unchanged. Demands for equity or justice are overlooked, with changes merely addressing consequences of change – not the problem itself, nor its causes.

Managerial Reform: Policies that show a greater commitment to rights, equality, and justice than *status quo* approaches but that seek to address these problems without challenging current structures or ways of doing things. Certain rules and norms might be altered or removed and new rules or structures created (for example, to increase access to skills and jobs), but the overarching structure of policy and intervention will not change.

Structural Reform: Policies that enact change based on allowing the greater participation of people and communities in the decisions made. The focus becomes on both the outcome and the form of intervention itself, ensuring a more inclusive process and allowing for further change and reform in the future. This might involve the public ownership of utilities or modified decision-making processes.

Transformative: The wholesale change of the structure, systems, and institutions that enable change, including those that are responsible for socioeconomic and climate crises. Rules are changed and modes of governance are altered to allow for alternative viewpoints and include processes to challenge and rewrite the status quo.

Many current regional, national, and international approaches to just transition overlap with the *status quo* and *managerial reform* interventions defined above (Pinker, 2020). Policies can often define a transition in the narrow, economic terms of employment, with limited effort made to provide alternatives or develop new regional economies. For example, the Latrobe Valley Authority, working in the coal fields of Victoria, Australia, created a Worker Transition Service with local employers to advise and assist those employees impacted by the closure of the Hazelwood Power Station. This focused on helping workers find new skills and training opportunities, financial support, and employment. Local branches of the national power industry unions developed worker transfer schemes to provide some jobs guarantee to those soon to be out of work (Snell, 2018). This is positive, ensuring employment for many but policies that prioritise returning workers to the workforce as quickly as possible can overlook the need to provide sustained support for communities undergoing wholesale change.

Status quo or managerial form interventions are mostly reactive, rather than proactive. They contain an implicit assumption that a transition is already happening and that influencing the form it takes and its impacts are impossible. The focus itself is on mitigating its consequences and supporting those affected, rather than seeking to define a more-positive and just transition (Huber, 2022). Policies that retain a primary focus on the plight of workers through jobs protections and re-training opportunities neglect the broader consequences of decarbonisation for communities – and how a restructuring of the economy and changes to employment has far-wider impacts. Sometimes plans can fail to detect these entirely. The Scottish Just Transition Commission (2020) reported that at a visit to Kincardine, the site of the recently closed Longannet power plant (the last coal-fired plant in Scotland), many argued that efforts to support those who lost jobs did well but that policy did not engage those living in the surrounding communities or understand how they had been affected. Yet, a just transition is about more than just workers in fossil fuel industries – it is about the families, communities, and broader regional economies around them. This has been addressed in the second report of the Scottish Just Transition Commission (2022), which argued that government policies must go beyond a narrow focus on 'green jobs' and promote policies that support everyone impacted.

The relatively slow pace at which just transition policies and principles have been enacted by the Scottish Government has led to dissent against the Scottish Just Transition Commission. It has been characterised by its critics as using of the language and framework of just transition without enacting meaningful change.[10] Friends of the Earth Scotland (2020) has claimed that the Scottish government has dismissed policies that hold transformative potential, such as calls for publicly owned energy and infrastructure companies and, instead, focused on limited measures. The organisation also highlighted the narrow membership of the Commission, including three advocates of the carbon capture and storage industry but little representation from trade unions, as likely influencing the policies that it might recommend (Dixon, 2021).

A just transition strategy can and should be proactive, seeking to alter and redefine what a transition is, and what workers and communities are transitioning towards. Such a move requires addressing who gets to influence such a process (Stevis and Felli, 2015; Huber, 2022). Any just transition will likely be complicated, at times contradictory and characterised by numerous tensions. To date, the term does still hold a narrow resonance. A survey of workers in the North Sea offshore oil and gas sector found that 91 per cent did not know the term 'just transition' or what it meant (Jeliazkov et al, 2020). This is a key constituency for decarbonisation in the UK, which should not only be aware of just transition narratives but have an active role in defining what they look like. For many, a 'just transition' is a narrative used by policy makers and academics only. It is necessary to explore not only what effective policies might look like but also how communities understand and influence their character, direction, and consequences. This also requires broader discussions of what the 'just' in a just energy transition is.

Energy justice, energy democracy and local transitions

With the definition of a just transition being flexible, it can be broadened to allow for decarbonisation that is inclusive for, between, and within different communities. Work is being done in universities, governments, think tanks, and beyond to rethink energy transition narratives to include demands for gender parity in decision-making, global climate justice, and the addressing of deep historical injustices (Webster and Shaw, 2020;

Lennon, 2021; Heffernan et al, 2022). A language of 'energy justice' provides a vocabulary to explore how access to safe, secure, affordable, and sustainable energy is often skewed on social, economic, and political lines. This is an important response to claims that decarbonisation is a technical pursuit by illuminating how it is often uneven and unfair across and within communities (Heffron and McCauley, 2014). Claims of energy justice and injustice exist along five key lines (Jenkins et al, 2016; Hazrati and Heffron, 2021). These are:

- **Distributive justice:** the distribution of costs and benefits of electricity generation and broader energy policy.
- **Procedural justice:** the inclusivity of decision-making processes in energy and decarbonisation policies.
- **Justice as recognition:** the recognition of the rights of different communities and groups and how they might be disregarded. Overlapping with procedural and distributive justice, *recognition* is centred on the need for respect, acknowledgement and the valuing of different identities, motivations, wishes, and outcomes.
- **Restorative justice:** in which policy corrects and repairs historic and harmful actions, such as responsibility for emissions or broader patterns of exclusion and inequality.
- **Cosmopolitan justice:** linking low-carbon transitions to a broader understanding of global injustice(s) and a discussion of 'embodied' injustices along international supply chains.

These forms of injustice overlap and coincide with one another. For example, energy poverty in the UK has been recognised as linked to the vulnerability of particular social groups. The elderly or those with Alzheimer's Disease and related dementias can often be more vulnerable to energy poverty (Heffron and McCauley, 2014; Jenkins et al, 2016; Liddell et al, 2016). Like other forms of poverty, energy vulnerability can also cluster in certain neighbourhoods – affecting those less financially able to weather rising prices or make changes to their homes (Bouzarovski and Simcock, 2017). While an energy transition may be formed of national policies, the experiences of energy injustice follow different geographies, becoming clustered in certain areas, and affecting particular communities disproportionately.

'Energy democracy' is an umbrella concept that has come to represent various calls by communities, trade unions, social movements, and others to ensure that decarbonisation retains includes the participation of people and communities in decision-making and a degree of democratic control over energy systems (Becker and Naumann, 2017). It is a complex term that simultaneously represents a goal of social movements or communities challenging energy policy decisions, a desired outcome of decarbonisation, and a process of decision-making itself (Szulecki and Overland, 2020). While this may be encouraged by top-down government policy, energy democracy has mostly been a bottom-up demand for broader change – in terms of how the energy sector is structured, how energy policy is made, and who has a say in both. This brings together different initiatives, led by residents or consumer groups, local and regional organisations, and public services (Routledge et al, 2018; Teron and Ekoh, 2018). Organisations, such as energy cooperatives, are key sites where energy democracy overlaps with notions of community organising and control and the potential of communities to control their resources (van Veelen and van der Horst, 2018). Owned and managed by members, these organisations provide an opportunity for local energy planning and decision-making to reflect members' priorities. These local-led responses to decarbonisation and calls for energy democracy are often where energy policies meet broader demands related to the control, finance, and ownership of energy. This can include a variety of calls, including the formation of energy cooperatives or calls for the return of the energy sector to public ownership from private hands (Angel, 2017; Allen et al, 2019).

It is hard to separate local discussions of energy projects from the connections that these residents and communities have to the neighbourhood, place, or landscape itself. In the new energy landscapes created by decarbonisation, these connections might be irreversibly altered. Think here of a favourite place that you might have. Places to sit and think. Sometimes, places to just sit and watch the world go by. We hold connections to the roads where we grew up, pubs and bars where we had our first drink, clifftops where we have sought solitude, towns where we fell in love, coastlines where we holidayed, regions that we hail from, and the destinations that we long to return to. Whist our connections to them may be messy and change over time, these places are never just points on a map. They are emotionally important to us and our experiences, both day-to-day and across key points of our lives (Tuan, 1977; Massey, 1991).

Such attachment to place evokes a 'rootedness', instilling a sense of belonging to a location, neighbourhood, and landscape (Devine-Wright and Howes, 2010). Connections are often individual, taking different forms for people within the same community. This could be linked to the length of time living in an area, insider/outsider status, and notions of rurality and disconnectedness from the busyness of cities (Stedman, 2002; Manzo and Perkins, 2006; Devine-Wright and Howes, 2010). All senses of place are deeply psychological, emotional, and social. They can highlight different experiences and inequalities in rural or urban environments (Bailey et al, 2016; Larkins, 2021; Windsong, 2021; Russell and Firestone, 2021). They are also inscribed with economic and historical context, with the concerns and perspectives of communities often rooted in their previous experiences and current expectations of infrastructure, transition, and change (Crowe and Li, 2020). Residents often link their opposition to renewable energy projects to these connections. This can lead to opposition to renewable energy projects being represented as 'not-in-my-backyard' (commonly shortened to NIMBY) style outcries, in which a group of residents are deemed to ignore the wider benefits of a project, being blinkered by the localised impacts. Up until the late-1980s, community opposition to schemes – be they roads, energy, or other types of projects – was perceived as NIMBYism.

A depiction of those opposing renewable energy projects as NIMBYs signals the underlying belief that communities might not have adequate knowledge to understand a project's benefits and that project decisions and planning should be left to the technical experts. This is characteristic of what has been labelled the 'information-deficit model' of policy-making, which understands public opposition to a policy as due to a lack of understanding or knowledge which can be addressed by giving more information and encouraging an apparent 'literacy' of the population (Sturgis and Allum, 2004). This approach might fail to make space for the variety of views, concerns, and fears held within a community.

A coherent, singular 'community' existing in one place is rare (Massey, 1991). Various perspectives can exist in the same place, complementing, contradicting, or failing to interact with one another. It is important not to treat a community at a given place as a unified bloc to be consulted and informed. We don't all have the same interests as our neighbours. Attitudes will differ, as will priorities. Some involved may prioritise low electricity prices, while others may stress energy security or prioritise the expansion of renewables (Hess, 2018). Similarly, resistance to renewable projects takes many forms. This includes concern about physical impacts but also more personal or sensory changes, such as the noise created by wind turbines and how this may impact everyday life (Deignan et al, 2013; Kim and Chung, 2019). Opposition can stem from actions that are protective of place, with residents highlighting the impacts of a project's siting and broader concerns about who has a say in the planning process, how and when (Devine-Wright, 2009; 2015). A sense of place can also lead to support for a project. For example, those living in coal communities may have a more positive response to new solar farms, due to their experience living at the sharp end of the fossil fuel regime (Crowe and Li, 2020).

Throughout my time spent writing this book, I have often returned to reflecting on what the label NIMBY means in an energy transition. The term is often used in a blanket way, thrown at wealthy homeowners contesting new house-building as much as local groups pushing back against infrastructure. This is often done with scorn, an accusation of NIMBYism is used to dismiss opposition and present their objections as only focused on protecting the local status quo (Dear, 1992; Galvin, 2018). Yet, local movements can also call for change and transformation or highlight a broader unfairness in how they are treated within policy. While it retains strength in some media reporting of such movements, it is important to move beyond thinking of opposition to projects NIMBYism and, instead, understand how opposition to renewable energy projects, such as at Graveney Marshes in Kent, is linked to the symbolic importance of locations (Devine-Wright and Howes, 2010).

At Graveney Marshes in Kent, many residents opposed to the solar farm highlighted its impacts on the landscape. One Faversham resident responded to the government consultation in 2019, voicing concern about 'the potential loss of this unique and beautiful area for future generations'.[11] This is not a rare concern about renewable energy infrastructure but the emotional bonds that residents have with the area and landscape can be overlooked in the planning processes that primarily focus on technical or economic details (Bailey et al, 2016). This exclusion can later lead to local opposition becoming inscribed with a broader political resonance. In Germany, the impacts of wind power on the landscape in Mecklenburg-Vorpommern gave rise to a new political party, *Freier Horizont* ('Free Horizon'), which participated in regional elections in 2016, arguing that wind energy was destroying the landscape and should be stopped (Liebe et al, 2017). In the UK, the UK Independence Party (UKIP) have previously presented wind farms as symbolic of the economic waste and political correctness of the European Union (Reed, 2016). UKIP resurrected this symbolism in 2022, claiming that wind turbines destroy landscapes, kill bats and birds, and increase energy bills (Nailer, 2022).

In the renewable energy landscapes brought by decarbonisation, residents and communities support and oppose new schemes by linking them to wider issues and narratives. These might include demands for participation, criticism of patterns of land ownership, or calls for the equal sharing of costs and benefits. All of which might interact with personal and collective connections to the places to be impacted. The links between a renewable energy scheme and these narratives can influence the legitimacy of the project or lead to it taking on a wider resonance (Nicholls, 2020). A just energy transition must not only recognise these place-based elements but elevate them and help communities use them to rearticulate the forms that energy transitions might take. While a just transition is rooted in the taking of national action to protect communities and groups at risk, it must also incorporate an approach of energy justice that gives the communities meaningful pathways to participate in an energy transition and its benefits. Such an approach highlights a significant disconnect in energy transitions: namely that the benefits of renewable energy infrastructure are tied to national targets and electricity supply but that the impacts are borne by communities at the local level.

3

Scale

The River Avon flows into Bristol from the southwest. Mallards and moorhens swim past and a grey heron stands on the bank. At dusk, bats might flit across the air above your head. The river forks as it enters the city. A large sign appears on the riverbank, directing all vessels to the right. Follow the sign this way and you find yourself in Netham Lock: a gateway to Bristol and its Floating Harbour. Beyond this, as buddleia washes out of cracks in the brick and concrete along Feeder Canal, you move past nightclubs and warehouses, timber merchants and car lots. You pass underneath the Meads Reach Bridge and, soon, you are in the city.

The New Cut runs to the other side of this fork. This waterway, built in the early 1800s, diverts the tidal Avon through south Bristol. The Cut flows past Temple Meads train station, Wapping Wharf and Spike Island. It takes in The Old City Gaol (now flats) and the Chocolate Path, joining the Cumberland Basin where another set of locks controls access to the harbour. There is much local folklore about the New Cut, including how it was built using the forced labour of French prisoners of war and that a party celebrating its completion ended with a riot.

While it has been possible for boats to navigate the New Cut in the past, this is only up to a point. This is what the sign at the fork warns against. The New Cut starts with a weir, a low head dam that manages the flow of water. Over 20 cubic metres of water pass over this weir every second (Bristol Energy Cooperative, 2020). The sound of this water washing over echoes across St Vincent's Industrial Estate, as kitchen fitters pick up supplies and mechanics complete their repairs. It is here that Bristol Energy Cooperative planned to build a micro-hydropower facility, installing two small turbines to produce enough electricity to power 260 homes and save around 520 tonnes of carbon dioxide emissions a year (Renewables First, 2019; Bristol Energy Cooperative, 2020). The Bristol Community Hydropower scheme at Netham Weir was in its planning stages between 2018 and 2022.

Like all of Bristol Energy Cooperative's projects, it would have been funded by the organisation's investor-members.[1] It was originally hoped (and planned) that additional funding would be found through the government Feed-in Tariffs (FiTs) scheme, in which payments are made to those generating energy that is sold to the national grid. The basic principle of feed-in-tariffs (discussed further in Chapter 5) was that a producer of electricity signs an agreement with an energy company or government that legally obliges the latter to buy a certain amount of electricity at an incentivised price for a set period. In the UK, this scheme closed in 2019 and, due to delays in obtaining planning approval, the Netham Hydropower project missed out with all the money available being spent within the set tariff band (based on the size of the turbine used).

This represented a setback, not just for this scheme but for many similar projects across the UK. The team at Bristol Energy Cooperative voiced a continued commitment to this project, planning to go ahead without government subsidies and securing £1.4m through community share offers and match funding from the European Regional Development Fund (Pipe, 2021).[2] However, in late 2022, it was announced that the community hydropower project at Netham Lock had been paused. Despite this setback, Bristol Energy Cooperative continued with other projects and, that same year, completed its biggest solar array yet – built on the city's Bottle Yard film studios in Whitchurch, with 2380 solar panels placed across the studio's rooftops. Funds raised by shareholders will help build this project, as well as rooftop solar arrays across the city and the formation of new microgrids at housing developments. A key reason given by many purchasing shares in Bristol Energy Cooperative to fund this work is the transparency of where their money goes and how it benefits the city. People supported this scheme because it is *theirs* and they can see where the benefits go.

In this chapter, I explore the scale of a just energy transition. The urgency of climate action and a shift to renewable energies has resulted in many schemes being built to provide as much electricity as possible. As I show in the case of hydropower, this can lead to numerous negative consequences that illuminate a distributive injustice of decarbonisation, through ecological destruction, the displacement of communities, and broader social consequences. In many places, the negative impacts of renewable energies are being pushed to the 'energy periphery', while the benefits (both in terms of energy security and economic returns) are enjoyed by communities elsewhere. This contrasts with smaller-scale energy projects, grounded in the local context, which can secure more benefits for local communities and landscapes. Whilst smaller energy projects do not provide the scale of change needed to fully decarbonise the electricity grid, they can enable communities to determine their own energy infrastructure: securing benefits and linking renewable energies to a sense of place and community.

The scale of energy transitions

The decarbonisation of the energy grid requires monumental shifts in how we use electricity and where we get it from. Staying within the 2015 Paris Agreement's global temperature rise limit of 1.5°C requires urgent, system-wide transformations of our energy grid now. This necessitates action by state governments – who have the influence, power, and resources to pursue such change and create the conditions to decarbonise. 8,000 GW of renewable energy generation is required by 2030 (UN, 2021). Globally, 260 GW was added in 2021 alone, despite a slowdown caused by the COVID-19 pandemic (IRENA, 2021a). This was mostly made up of solar and wind energy, which now represent over 50 per cent of all renewable energy capacity globally (IRENA, 2022).

There is often a 'scalar bias' in energy infrastructure. Energy planners prefer large utility-scale facilities over small-scale deployments as they have the capacity to generate more electricity and provide for the grid (Sareen and Haarstad, 2021). Vast offshore wind farms have hundreds of turbines. The biggest, the Jiuquan Wind Power Base in Gansu province, China may soon have as many as 7,000 turbines. Expansive solar farms spread to the horizon. The Bhadla Solar Park in Rajasthan, India, extends over 57 km². The scale of this infrastructure represents a more direct route for many energy planners to decarbonise electricity grids than smaller, localised moves to expand smaller-scale energy infrastructure or put solar panels on the roofs of individual buildings.

Schemes of this size require vast financial investments, extensive supply chains, and concerted collaboration between powerful energy players. With big energy infrastructure required, national governments work with large energy companies that can raise the funds and provide the expertise needed for new wind farms and solar installations to be built quickly. For many of those investing in such schemes, the desire for profit requires quicker financial returns. Once built, large scale renewable projects can provide this. The declining cost of solar has allowed utility-scale projects to grow larger and larger. Due to the lower costs of the technology, the only limit now placed on the size of a solar farm is the land available for projects to be built upon. As the expansion of solar facilities continues across landscapes in many countries and regions, it is possible that solar energy generation may occupy up to 5 per cent of total land (van de Ven et al, 2021). These sites are often fenced off, breaking up the landscape, habitats, or migratory pathways (Mulvaney, 2022). Their construction scrapes the soil and floods an area with solar panels, obscuring the ground and creating new energy landscapes. As solar energy generation expands, these facilities will alter future patterns of farming, land ownership, and rural life.

The siting of large scale renewable infrastructure is often unevenly spread across regions and communities. By their very nature, large energy projects must be built in rural areas – where there are fewer people to be impacted and/or a greater availability of land to build on. These are the 'energy peripheries', areas that are becoming home to renewable energy projects but exist at the edges of political and economic power with limited access to opportunities and services (O'Sullivan et al, 2020). Communities at this periphery are disadvantaged by the whole energy system. Energy networks are denser in areas of high demand, such as in cities, but sporadic elsewhere. The expansion of renewable energy infrastructure will further illuminate this unevenness. Burdens associated with renewable energy might include air or water pollution, the loss of lands or sites of cultural importance, and damage to habitats and areas of rich biodiversity (Levenda et al, 2021). These impacts are often project-specific but represent a process of change and transformation of a region in ways that may trigger wider opposition.

An important way to understand energy peripheries is through a language of *distributive injustice*. Taken broadly, distributive justice focuses on the lived experience of how harms and benefits are distributed and experienced by people and communities (Kaswan, 2021). This forms a core part of broader environmental justice scholarship that has evolved over the past four decades to detail how environmental burdens and impacts disproportionately affect poor or marginalised communities or Black communities[3] (Bullard, 1993; Pulido, 2000). The initial impetus for the environmental justice movement in the early 1980s in the United States was the exposure of and opposition to the siting of toxic waste facilities, found to be disproportionately close to Black communities. In the decades since, environmental justice has become a central organising narrative around which local, national, and global environmental and climate movements are formed. These movements have developed into a rich collective that incorporates numerous understandings of justice and demonstrates how pollution and the impacts and benefits of new projects are rarely evenly distributed across groups and can be skewed by race and ethnicity, class, income, gender, and age (Schlosberg and Carruthers, 2010; Bell, 2014).

An approach of distributive justice helps us understand the tension between the local impacts of renewable energy projects and how energy policy is made at the national level. The planning, funding, and installation of renewable energy projects at the energy periphery can often create conflicts between planners in central government and those living in the regions impacted. Within such tensions, the protection of local ecologies, landscapes, and residents becomes placed in direct conflict with broader national ambitions of decarbonisation and climate action. In many countries, these contests represent a continuation of historical patterns of marginalisation and unfairness. In Norway and Sweden, the construction of wind energy projects in the Sápmi (popularly known as Lapland) has represented a renewal of historical processes of the appropriation of land and the displacement of the Indigenous Sámi communities (Normann, 2021). The construction of the

Markbydgen and Øyfjellet onshore wind farms in northern Sweden and Norway respectively are to be built in the traditional migration paths of the reindeer that the Saami are reliant upon for their livelihoods (Lawrence, 2014; Muotka, 2020).

Globally, this has taken the form of a new 'green industrialisation', in which facilities are built with limited concerns for local communities or ecological impacts (Brock et al, 2021). Large solar farms have become part of broader patterns of exclusion, land-grabbing, and profiteering. The world's largest solar park – found in Gujarat, India – has increased the precariousness of already-vulnerable communities due to a loss of community-held lands and livelihoods (Yenneti, 2016). In Kerala, India, the expansion of large solar utilities overlaps with local and regional land tensions, with land taken from poorer Adivasi households and communities who did not hold formal land rights without compensation (Bedi, 2019). On the Isthmus of Tehuantepec, Mexico, wind farms have been proposed to ensure supply to large corporate, multinational consumers of electricity, such as Coca-Cola and Walmart, despite localised impacts and opposition (Bessi and Navarro, 2016). In this region, wind farms have exacerbated social inequalities by only making compensation payments to landowners, rather than providing some form of community benefit payments, and neglecting the impacts faced by those who do not hold formal land tenure (Avila-Calero, 2017).

In all cases, the scale of energy transitions has resulted in a cost-shifting of impacts to the energy periphery in the name of national emissions reduction targets. The imbalance of benefits and impacts has been presented as resulting in what are described as 'green sacrifice zones' (Zogrofos and Robbins, 2020). The term 'sacrifice zone' stems from areas made uninhabitable by the fallout of nuclear tests during the Cold War, with these areas *sacrificed* for the national security priorities of the term. It has provided an important vocabulary to describe places where residents faced disproportionate exposure to risk and toxicity from pollution in the name of national priorities – with this imbalance overlapping with other patterns of discrimination against and exclusion of Black communities (Lerner, 2010). Regions impacted by previous energy projects can again lose out in today's energy transitions. For example, in South Catalonia, Spain, wind power promised to bring a better, more-local energy system but the sector has continued to be dominated by the same companies (Franquesa, 2018). Wind energy has become just the next energy source that props up these interests and, in doing so, further has cemented South Catalonia's place as an energy periphery. The region, historically home to nuclear power, oil refineries and hydroelectric dams, has entered a new era in its history as a renewable energy landscape – yet the winners remain the same.

The language of 'green sacrifice zones' is used by those critically discussing renewable energy projects (Scott and Smith, 2017; Castán Broto and Calvert, 2020; Brock et al, 2021). It is useful for understanding the broader concentration of impacts of decarbonisation and renewable energy infrastructure, and how the groups who faced these impacts often remain excluded from its benefits. While the scale of renewable energy infrastructure is linked to national targets for decarbonisation and patterns of investment, the impacts are always experienced locally. These localised impacts become linked to wider patterns of unfairness and distributive injustice, with the imbalance of benefits and impacts often seen in those who financial invest in and benefit from new renewable energy infrastructure. This has been particularly evident in the role of hydropower in contemporary energy transitions.

The distributive injustice of hydropower

The presence of 'green sacrifice zones' at the energy periphery occurs within a context in which renewable energy projects represent new streams of global investment, revenues, and profits – all while placing new demands on energy landscapes (McCarthy, 2015; Spivey, 2020). With the decarbonisation of the energy grid requiring extensive financial investment, numerous avenues of international and regional funding have sprung up. In the case of hydropower, this has taken the form of financial mechanisms that place energy landscapes within a broader web of international finance, global emissions targets, and, at times, localised corruption.

When we think of a dam, we conjure an image of a bulk of earth or rock or concrete blocking a river. These are broadly embankment dams (built from earth or rock) or gravity and arch dams (from concrete). All create large reservoirs to store water, releasing it through turbines to generate electricity. The 20th century saw a vast expansion of hydroelectricity, often paid for by public funds from national governments or assistance from international organisations, such as the World Bank (Everard, 2013). Towards the end of the century, international institutions reduced their role in the direct financing of dams – prompted by growing criticism of the social and environmental impacts of hydropower projects (McCully, 2001; Scudder, 2006; World Commission on Dams, 2000).[4]

Despite knowledge of these impacts, hydroelectric capacity doubled between 1995 and 2020, extending to over 1,300 GW (IHA, 2020). In 2020, hydroelectric dams generated 4,370 terawatt-hours (TWh) of electricity globally, roughly equal to the total energy demand of the United States (IHA, 2021a). In 2021, new hydropower projects that generated 21 GW of electricity were put into operation, including projects in Canada, Laos, Nepal, India, Ukraine, and China (IHA, 2022). Hydropower remains part of the energy and decarbonisation plans of many countries, such as China, Turkey, Peru, Laos, and Myanmar. This expansion has been justified as providing energy security to support growing populations and expanding economies.

Considering the vast scale of change required by any energy transition, hydropower is often presented as a route to reducing society's dependence upon fossil fuels energies (IHA, 2021b). It holds an important utility in decarbonised energy grids. Proponents of hydropower argue that it can store and provide a secure, base load of energy that can be released when other energy sources, like wind and solar, are unable to fulfil high levels of demand (IEA., 2021b). The infrastructure can act as a 'battery', providing a ready pool of electricity when needed. This also involves calls for expanding pumped storage hydropower capacity – in which water is pumped water to a higher reservoir before releasing it down a pipe to a lower reservoir, running the water through a turbine to create electricity. However, many hydroelectric projects remain schemes that block rivers, create reservoirs, and run water through turbines.

Claims of the sustainability of hydropower have been supported by the development of the Hydropower Sustainability Assessment Protocol (HSAP) by the International Hydropower Association to assess the sustainability of hydroelectric projects and report best practices. The HSAP, developed between 2007 and 2010, provides 24 criteria (ranging from climate change mitigation and resilience to population resettlement), that can be used to gauge best practices and the effectiveness of project planning and development. The protocol has now been adopted by funders, such as the World Bank, Societe Generale, and Standard Chartered, to analyse and support the development of 'sustainable' hydropower.[5]

Hydroelectric projects are an important example of the scale of renewable energy transitions, and their distributive justice implications. The nature of large dams being expansive projects of concrete and blocked rivers necessitate their construction in rural areas, away from population centres. The environmental impacts of hydropower include changing river flow and hydrology, water levels and quality, disrupted fish migration pathways, impacts biodiversity, altered sediment flows, and the emission of greenhouse gases, such as methane, due to the decomposition of trees and other vegetation underwater.[6]

The most dramatic social impacts of hydropower are seen in how people are forced from their homes to make way for the project, which disrupts the lives and livelihoods of those affected, often for generations (Scudder, 2005; Takesada, 2009). An estimated 80 million people have been displaced by hydropower projects to date (IDMC, 2017). Displacement takes the form of lost land, houses, and neighbourhoods and the disruption of social relationships. Those displaced may also be paid inadequate compensation, or the homes where they are rehoused may fail to be up to standard, resulting in further issues due to a lack of support after they have been resettled elsewhere (Weißermel, 2021). Even within more comprehensive resettlement programmes, it may take generations for families and communities to recover.

When defining the apparent the 'sustainability' of hydropower, there remains a real neglect of the cumulative impacts of hydropower on hydrology and ecosystems. The wholesale damming of rivers leads to dramatic impacts that are experienced upstream and downstream (Winemiller et al, 2015). Impacts occur across landscapes, with hydroelectric projects affecting communities both upstream (for example, through restrictions on water use) and downstream (through reduced river flow). The social impacts of hydropower projects don't just take the form of displacement but can also include knock-on consequences that change the lives and livelihoods of many. Health impacts include those caused by vectors thriving on a new reservoir's surface and linked to increased levels of methylmercury in the water (caused by the flooding of soils). Hydroelectric projects are also disproportionately built on Indigenous peoples' land (Cooke et al, 2017; Hendriks et al, 2017). This has important cultural and territorial impacts, through encroaching on protected territories, flooding sites of cultural importance, and leading to further negative impacts. All of this can have important consequences for health, security, and gender relations in the areas. The land is flooded and lost, livelihoods are altered, and changing hydrology creates uncertainty for many. It is due to these impacts that the 'green' credentials of hydropower, of all designs and sizes, have been increasingly challenged (Ahlers et al, 2015; Atkins, 2020).

These impacts have often occurred in the name of global climate action. Financing hydropower in renewable energy transitions in many places has come from the Clean Development Mechanism (CDM). The CDM was set up, as part of the 1997 Kyoto Protocol, by the United Nations in 2001. Its goal is to channel financial investment from polluters and emitters in the Global North to fund 'green' projects in the Global South. In essence, it is a carbon offset scheme – with countries able to fund emissions-reducing schemes in other countries and claim these savings as part of their efforts at climate action.[7] Hydroelectric projects funded this way include the Barro Blanco project in Panama, which had dramatic impacts on local communities, including the displacement of Indigenous Ngäbe families (CIEL, 2016). In Uganda, the Bujagali dam inundated the cascading rapids of the same name, a site of cultural and spiritual importance for local communities (International Rivers, nd.). Both projects received funding due to their aiding of global emissions reductions – with residents facing significant impacts in the name of international benefits.

While many CDM-funded projects might fulfil the objectives of emissions reductions for the donor country, very few contribute positively to sustainable development agendas in the host country (Sutter and Parreño, 2007). The loss and trauma experienced by communities has allowed for reported carbon 'savings' and the offsetting of responsibility and emissions for richer groups in the Global North. Local communities lose out in a trade-off between international priorities and local benefits.

Investment in CDM projects has become less common in the past decade, due to a declining market price of carbon credits and global financial uncertainty. However, hydropower remains an appealing investment to many – due to it being a cost-competitive source of electricity (providing cheap, reliable energy) and an important part of decarbonisation agendas. Recent years have witnessed a rush of investors into hydropower and other renewables, seeing them as a relatively safe investment that guarantees future returns (Merme et al, 2014). This network of funders includes national banks, financial institutions, and energy companies, all of whom have seen potential for new investments and secure profits.

A rush to invest in hydropower has been encouraged by governments, who change regulation, overlook impacts, and provide opportunities for profit. This web of finance is evident in the expansion of smaller dams in south-eastern Europe – which, at the time of writing, is one of the central areas of the expansion of hydropower. An important driver of this expansion is the presence of direct financial support, in the form of subsidies implemented by national governments to stimulate investment in renewable energies (to meet European Union decarbonisation targets). In 2018, 70 per cent of the subsidies available in this region were provided to small hydroelectric projects, despite hydropower representing only 3.6 per cent of total electricity generation (Bankwatch, 2019). In Bosnia and Herzegovina, smaller dams, many privately owned by individuals, have provided fertile ground for those seeking to profit from an energy transition. Owners have been accused of manipulating funding processes, bribing officials, and obtaining planning approval while neglecting or overlooking impacts (Dogmus and Nielsen, 2019). Importantly, many small dam projects have been approved but remain unbuilt. In 2017, only 6 per cent of the hydropower projects planned in Bosnia and Herzegovina were actually under construction (Dogmus and Nielsen, 2020).[8]

Incentives for renewable energies in the regions hosting these dam projects have not been made accessible to people and communities. Instead, they have been primarily used by and benefited those who are already wealthy and powerful. Bankwatch (2019), a network that monitors public-financed projects, has reported that: in Montenegro, the system has primarily benefited those close to President Milo Đukanović (1998–03; 2018–) and, that a key beneficiary in Serbia was the best man at President Aleksandar Vučić's (2017–) wedding. In North Macedonia, the Deputy Prime Minister, Kocho Angjushev (2017-) is understood to have owned 27 small hydropower facilities (Bankwatch, 2019). The leader of the opposition, Hristijan Michoski owns at least five (Bankwatch, 2019). This represents a monopolisation of the financial incentives to build hydropower, with powerful, well-connected small investors using subsidies to guarantee profits and returns.[9] Locally, this comes to represent the enclosure of the river by the rich and powerful who are seeking to personally profit from energy transitions.

While the Hydropower Sustainability Assessment Protocol has presented hydroelectric projects as 'green' and 'clean' energy generators, the energy source holds a continued distributive injustice. Renewable energy sites create an energy landscape of differentiated risks and benefits. The costs of decarbonisation are shifted to the energy periphery, with regions elsewhere 'exporting' the impacts associated with their energy needs and demand to communities elsewhere. In light of the extensive impacts of hydropower, this distributive injustice involves the transformation of rivers, with water resources channelled and redirected to new interests and the local communities reliant upon the waterscape left to face the impacts.

Local transitions

The technologies linked to decarbonisation are 'scalable'. They can be deployed and used in various places and at various sizes. Solar energy can be deployed in expansive arrays that cover fields as far as the eye can see, or it can be used on the roof of somebody's home. Hydropower can also be deployed at the local level, with smaller facilities being used to both generate electricity but also to instil a sense of local ownership and action in an energy transition. While smaller hydropower projects have provided a terrain for some to seek new profits, the technology can also be owned and deployed by local community organisations. Across the UK, community energy groups are building micro-hydro plants on weirs and rivers – from the River Calder at Whalley, Lancashire to BroomPower in Ullapool, Scotland. Investors are primarily small donors from the local area. For the Osney Lock Hydro project in Oxfordshire, two-fifths of investors lived within a mile of the project.[10] The project now provides enough electricity to the national grid to power more than 500 homes. These local responses to energy infrastructure cannot be understood using a vocabulary of NIMBYism only. Instead, they are fruitful sites for the emergence of alternative visions of energy transitions.

Communities often support renewable energy projects being built in their area but with important conditions. Recent work has found that local communities in Rhode Island in the United States support offshore wind farms with one important condition: that it does not go to a nearby 'rival' state and population (Bidwell et al, 2022). This is a notion of 'regionalism', in which communities don't want to see a landscape used for electricity that is primarily for communities elsewhere. In some cases, residents may welcome new energy projects due to this regionalism and its connection to past experience of hosting energy infrastructure and a first-hand knowledge of the damage brought by fossil fuels (Van der Horst and Evans, 2010).

Those opposed to energy projects don't necessarily discuss them using the language of distributive justice discussed above. Instead, their responses may become linked to calls related to economic or community impacts, or changes to the local landscape (Bailey and Darkal, 2018). This can be tied to the scale of infrastructure: smaller-scale or rooftop solar is often more popular than larger facilities (Nilson and Stedman, 2021). Small community-centred hydropower can transform economies and reconnect people to energy generation and do so without the expansive impacts associated with bigger projects. International Rivers (2022), an organisation that stands against big hydropower across the globe, has highlighted the importance of these smaller energy systems in rural areas. At the energy periphery, small-scale energy generation can allow for autonomy and self-sufficiency in electricity use, with the local community able to generate their own electricity and enjoy the benefits rather than witness the landscape altered to provide it for those living elsewhere.[11]

Community-centred stories of transition are formed of broad webs of people and organisations simultaneously acting individually and together. A commonly used example is found in the Orkney Isles – an archipelago formed of 70 islands, 20 of which are inhabited by a shared population of just over 22,000, lying to the northeast of Scotland. The electricity needs of Orkney are mostly fulfilled by both larger blocks of wind turbines and smaller, household-owned turbines. The islands now generate more electricity than their annual energy demand. Orkney's geography has made it an ideal location for many organisations working at the forefront of wind and tidal energy. Waters off the coast are home to new schemes testing wave and tidal power, with the archipelago home to the European Marine Energy Centre (EMEC), set up in 2003 and acting as a key testing site for the technology.[12] The presence of EMEC provided a new impetus to the regional economy, attracting new organisations to Orkney. These include industrial and consultancy groups (such as Aquatera, which provides expert support in the planning of renewable energy projects) and research organisations (Edinburgh's Heriot-Watt University operates its Orkney Research and Innovation Campus in Stromness). The arrival of each organisation brought new jobs and skills, and gave Orkney even greater visibility in international energy transitions.[13]

The strength of Orkney's energy transition comes from within. The community retains a strong stake in the renewable energy future of the islands. Innovative energy ventures remain community-run – such as Surf 'n' Turf, which is led by Community Energy Scotland and aims to generate hydrogen from tidal and wind energy. The Orkney Islands Council (OIC) and Orkney Renewable Energy Forum (OREF), both local-led organisations, retain important roles. OIC has teamed up with EMEC and other local partners on the £285 million ReFLEX project that aims to use digital technologies to create a 'smart' energy system that monitors and controls technologies to ensure supply meets demand. This emergent energy source is presented by the OIC as a key route to addressing energy poverty on the islands, where old houses, low incomes, and a high cost of heating led to 57 per cent of households living in energy poverty (OIC, 2017). OREF provides an umbrella for residents in their work with energy organisations and, across Orkney, these groups regularly work together – even when, elsewhere, they might be competing or standing in each other's way.

The fact that the Orkney Isles is an archipelago is important. Island energy systems are sensitive to both their natural surroundings and events elsewhere. Like the Shetland Isles to its north, Orkney's historical links to the extraction of offshore oil and gas in the North Sea had led to a reliance on accommodating markets on the mainland and vulnerability to change patterns of supply and demand (Johnson et al, 2013). This island-ness has also given rise to a sense of place, togetherness and autonomy, with wind energy providing a route to ensuring that the island communities become self-reliant as well as avoid being left behind by changes elsewhere.[14] These relationships, and their links to renewable energy infrastructure, aren't limited to the present but, instead, have deep roots in the past and extend into the future, in terms of geography, patterns of land ownership, and identity (Watts, 2018). Orkney has been a key site in the history of wind energy in the UK, with the first turbine in the country sited at Costa Head in 1951 (Dudley, 2020). Today, renewable energy provides a force to mobilise residents. The proliferation of domestic wind energy has given rise to new levels of energy literacy, not only of how electricity is used but also through a knowledge of infrastructure, energy prices, connections, and wind as a resource itself.[15]

With decarbonisation an important part of national economic development and climate action agendas, the experience of Orkney demonstrates the different forms that renewable energy infrastructure might take. The case of community-centred energy in Orkney highlights how decarbonisation can be place-based – becoming linked to the landscapes and physical characteristics of area region, local institutional arrangements and residents' connections, and sense of place. Across Orkney, renewable energy is as much about the people and community as it is about the technologies tested, deployed, and operated (Watts, 2018; Ford, 2022). Orkney and those who live there (known as 'Orcadians') have been presented as holding different identities from those living on the mainland – enjoying an important sense of place, community, and interconnectedness, both within communities and across the archipelago itself (Watts, 2018). Renewables have become part of Orkney's 'brand' and sense of place, with students, journalists and other interested people travelling to the islands to learn more about renewables. Hotel roofs hold solar panels and local business owners show an awareness of the renewable energy projects nearby. For a generation of residents, renewables are not only the future, but they are also part of the present. As one employee at the European Marine Research Centre, explained to me: the industry had grown over their lifetime and they can remember certain projects, like the Burgar Hill wind farm, being built as if they were local landmarks. The Orcadian energy transition highlights the important role of place-based decarbonisation, and how renewable energy companies can work with residents and local groups to generate a more inclusive, community-centred energy transition.

Rule 1: Push for community-centred renewables

The location of hydroelectricity in contemporary discussions of decarbonisation neglects an important distributive injustice – of projects being built at the energy periphery, rivers becoming batteries, and the benefits being transmitted and enjoyed elsewhere. The energy landscapes created represent the displacement of costs, risk, and impacts – with local communities losing out in a trade-off between their lives and livelihoods, on the one side, and national policies and international investments, on the other. The hydroelectric project at Netham Weir in Bristol that I described at the beginning of this chapter is incredibly small when compared to large dams. The Three Gorges dam in China, the world's biggest hydroelectric project, is 100,000 times bigger. Yet, Netham Weir represents an important route forward: a community-centred project that seeks to minimise impacts and provide a new space for locally owned electricity to be generated.[16] Like the Bristol Energy Cooperative, many local organisations are funded by share offerings, in which members buy a share in the cooperative, gaining a financial stake in the projects, investing in renewables, and having the opportunity to influence future work. Decisions are based on a principle of one member-one vote – rather than on the number of shares owned, which would prioritise the wishes of the wealthiest. Investors can expect some interest payment on their investment but these aren't particularly high. However, profit often isn't on the minds of these shareholders, with organisations investing the revenues into good causes, community funds, or to support energy projects elsewhere.[17]

Current energy transitions are underpinned by large energy projects being built at scale – transforming landscapes but with limited benefits for the local community who are experiencing such change on their doorstep. This large scale can obscure how communities might play their own role in energy transitions. Community-owned hydropower in the UK doesn't necessarily address a distributive injustice in energy transitions. However, I was struck by the strength of the support that this project had received from residents in Bristol. Looking at the comments on crowdfunding pages, the support seems rooted in two things. First, the project is led by a local organisation. Second, donors could see where the money comes from, where their investment goes – and that it will benefit the local community. Who owned and managed the infrastructure mattered.

Decarbonisation can provide an umbrella for new relationships between communities, businesses, institutions, and those organisations working on renewable energy technologies. State financial support, both from national and regional governments has often been a key driver of community energy schemes. In Orkney, the expansion of hydrogen infrastructure has been partly funded by the Scottish government and the Highlands and Islands Enterprise, a regional development agency (Duff, 2022). Similarly, the UK government release a £15 million Rural Community Energy Fund in 2013 to support rural communities exploring developing their own renewable schemes. In 2017, the government followed this scheme with the creation of five regional hubs (now called 'Net Zero Hubs') in England that aim to support local communities and organisations to develop renewable energy schemes and secure financial support for them. The Scottish Government has also aimed to create similar Community Climate Action Hubs with similar remits. These can provide important points at which local energy projects and national policy aspirations can be bridged and support one another (Hannon et al, 2022).

For a just energy transition, financial support should be readily available to communities seeking to develop their own energy infrastructure and direct a local energy transition. Financial grants might allow for local-level, small-scale experimentation. The provision of low-interest loans or the provision of loan guarantees to other lenders can help build infrastructure, advancing the money required with it to be paid off once the community starts benefiting (Hannon et al, 2022). Further support for communities can come in the national government leading processes that help locate any energy transition at the local level, providing key spaces for the sharing of skills and knowledge and actively seeking to match-make between communities and organisations working in the renewable energy sector. Groups seeking to develop their own community energy schemes can also be hindered by a lack of access to finance and expertise other communities enjoy (Bird and Barnes, 2014; Lacey-Barnacle and Bird, 2018). Work is needed to expand the arenas available to help schemes get off the ground and to support those interested in developing the skills and knowledge required for a successful community-run scheme. There is a need for local schemes to collaborate with intermediaries in community energy – different groups who, while not local to the area, share technical expertise, knowledge of finance and policy, and lessons learnt from elsewhere.

To support intermediaries and the sharing of knowledge and skills, Hannon et al (2022) have called for the creation of a new national, independent community energy delivery body that would provide support for communities in skills training, managing financial, legal, or regulatory hurdles, and working with others to learn about what works (and what doesn't). Such an organisation would be essential, allowing for the building of links between communities, decision-making institutions, and renewable energy companies – similar to the collaborations that have driven the success of Orkney's energy transition.

Energy justice is about more than the fair distribution of benefits and burdens. It must also involve communities having a greater say over the infrastructure built, how it is used, and who benefits. Addressing the distributive injustices of contemporary energy transitions, in which impacts are pushed to the energy periphery, requires a greater focus on the scalability of renewable energies. While large-scale projects may represent rapid, wholesale emissions savings, they can also form new energy sacrifice zones and displace communities, in the name of global flows of financial investment and climate action.

The Orkney model brings together renewable energy, the community and its institutions, and the local economy to harness abundant energy resources in a way that not only innovates but provides for the community. It reaffirms the links between the community and energy resources and develops new connections between local organisations and those specialist research institutes that develop emergent technologies and ideas. Community support and action have been harnessed and nurtured to create an energy transition that is particular to Orkney. Similar transitions can happen elsewhere. A just energy transition must seek to empower and enable community-based renewable energy projects through policies that prioritise the development of these schemes by creating incentives and financial and regulatory support for communities interested in following Orkney's lead. This could include policy support to ensure that energy resources are used to meet local needs, by setting up new institutions (such as the Orkney Renewable Energy Forum) that define the direction and pace of any transition. The skills gained from such a process can then be scaled-up to build new infrastructure and industries in the region and support transitions elsewhere.

The experience of Orkney's energy transitions highlights the potential of a just energy transition, in which the fair distribution of costs and benefits is contingent upon other forms of justice and fairness. While significant, only focusing on distributive justice may risk overlooking the political processes that underpin such inequities and that might allow for transformation (Foster, 1998). This raises a question of procedural justice – in which decision-making processes in energy and decarbonisation policies are inclusive and open to all.

4

Ownership

We speed past sailboats as spray comes over the side, holding on with both hands and staring into the distance. The Sussex Downs meet the coastline behind us, as the cliff faces of the Seven Sisters fade into the sea. All soon becomes haze and horizon. The engine slows, the wind calms and the sun beams down. Giants tower over us, extending in their rows, turning in time. White towers standing on bright-yellow bases. Formed of fibreglass and carbon fibre, their blades bend with the wind. Catching it, putting it to use. There is little sound, with no cry of gulls or calls from the seaside. Just the hum of the engine and the turn of a blade.

The Rampion offshore project spreads out over 70 km². Its 116 80-metre-high wind turbines are organised into 12 rows and joined by over 27 km of buried undersea cables. The turn of these blades drives a generator in the nacelle, the small shell found at the top of the tower. The electricity generated is directed to an offshore substation. As we float towards it, we can see where the cables join the substation in large, yellow numbered tubes. Identical pipework holds other cables that journey to the shore. They make ground at Lancing, close to the beach where I often sat as a child and looked out to what was then an empty horizon. From here, the electricity moves inland through buried transmission lines beneath the South Downs, joining the national grid at Twineham and powering 350,000 homes.

Rampion was the first offshore wind farm built on the south coast of England. The project, known as the Southern Array, when approved in 2014, was given the 'Rampion' name by a winning entry to a competition for local schoolchildren. The name refers to the county flower of Sussex, a sharp-blue bloom found on the Sussex Downs. Our guide on this trip told us that the winning entry received a prize of £100 for naming a project that, when completed, cost close to £1.3 billion.

The turbines have become a recognisable landmark on this stretch of the coast. Telescopes on numerous beaches give you a better view. A visitor centre is found on Brighton Seafront. Beneath the waves, the restriction of trawling activities around the towers has encouraged the return of marine life and rich ecosystems. This is just the beginning. RWE, the German multinational energy company that operates Rampion, has recently announced Rampion 2, whose turbines may be over twice as tall, extending up to 325 metres high at the top of the blade.

As I moved offshore in March 2022, the formal consultation for this new project is open. Across the coast, people received letters and leaflets advising on what the next project looks like and requesting their responses. Several had already spoken in opposition. The Sussex Wildlife Trust (2021) voiced concern over the proximity between Rampion 2 and marine conservation zones. In local media, a member of the Middleton-on-Sea Coastal Alliance (MOSCA), which publicly opposes the project, argued:

> It doesn't take much imagination to appreciate the visual impact the installation of 116 turbines up to 325m tall topped by flashing red beacon lights will have for residents and visitors, both from the shoreline and the South Downs National Park ... We welcome wind farms that produce truly green energy and having studied the proposals have concluded the proposed Rampion 2 wind farm does not. (Smitherman, 2022)

Opposition to wind energy in the UK has been influential at the national level, particularly in restricting the expansion of onshore wind. In 2015, Amber Rudd MP, the then Secretary of State for Energy and Climate Change ended subsidies for onshore wind. While the official logic behind this move was the need to invest in other technologies, it also came from pressure from those concerned that the expansion of wind would damage valued rural landscapes. The removal of subsidies was accompanied by the restriction of onshore wind projects in two ways. First, they could only be built in areas officially designated by local authorities. Second, to be built, a project must address all concerns and impacts identified by local communities. There are claims that this was not an official moratorium, but it has been a major brake on onshore wind projects. Only 11 onshore schemes, involving 20 turbines with a total maximum capacity of 42 MW, were approved in the UK between 2016 and 2021 (Windemer, 2022). This is a mere 2.7 per cent of the number granted in the years before the decision – with 157 wind farms, 730 turbines and 1.6 GW of combined capacity approved between 2009 and 2014 (Windemer, 2022).[1]

This is an important moment for wind energy generation in the UK. More regions, both onshore and offshore, may be altered and transformed into renewable energy landscapes and more host communities may question or oppose such change. In 2021, the Crown Estate auctioned off rights to the seabed for six new offshore wind farms in England and Wales with a potential generation capacity of almost 8 GW (Crown Estate, 2021). In January 2022, leases to 17 new offshore projects with a capacity of almost 25 GW were also awarded (Crown Estate Scotland, 2022). The energy crisis of 2021–2022 prompted government pledges to further accelerate this roll-out of offshore wind energy. The release of the UK Energy Security Strategy in April 2022 provided further evidence of a focus on offshore, rather than onshore wind. The strategy raised the 2030 target of offshore wind generation to 50 GW, signalling a steep increase. Despite regular and repeated discussions of easing the effective ban on onshore wind, it was noticeably absent in this strategy, with an initial target of 30 GW seemingly removed due to concerns about local consent and acceptance (Vaughan and Waugh, 2022). It has continued to be criticised by opponents, including those in government: in April 2022, the then Transport Secretary Grant Shapps labelled onshore wind technology an 'eyesore on the hills' (Merrick, 2022).

In this chapter, I explore how both onshore and offshore wind energy projects focus on securing the acceptance of host communities impacted by the project through public consultations, the payment of compensation, or community benefit schemes. However, these attempts can highlight a wider procedural injustice of renewable energy transitions: in which host communities are treated as merely passive observers of electricity generation and recipients of reimbursement. There is a need and an opportunity to rethink how wind power can be linked to wider questions of ownership of both land and energy, and how communities might be given a more active role in defining wind's place in a just energy transition.

Wind energy and renewable energy landscapes

The turbines of offshore and onshore wind are essential parts of decarbonisation. They also illustrate the significance of the landscape in a just energy transition. Large in size, they require vast landscapes to maximise the amount of wind that can be harnessed. The infrastructure is fixed in place and, due to the need for regular winds, it is often found spread out across areas that are highly valued for their scenic qualities such as upland areas and marine landscapes offshore (Warren and Birnie, 2009). It is here that wind turbines create new energy landscapes, transforming coastlines and rural scenery. Just as Rampion 1 transformed the Sussex coastline, Rampion 2 will further alter how shoreline communities interact with the coastal landscape. These turbines stand ever taller: the structures of Rampion 2 are expected to be as tall as the Eiffel Tower. It is near-impossible to make them invisible and ignorable, even when they are out at sea.

The arrival of wind turbines and transmission lines have been seen by some host communities as incursions that ruin a landscape. Common concerns about wind power include risks to biodiversity and habitats and threats to marine or flying wildlife, such as bats and birds. The disruption to the landscape is a prevalent concern for host communities affected by onshore and offshore wind projects and is important in understanding why residents might oppose a wind project (Roddis et al, 2018). This may be influenced by many physical aspects: such as the size of the turbines, the total footprint of the wind farm itself, its closeness to houses or communities, or where it is sited (Meyerhoff et al, 2010; Ek and Persson, 2014; Lienhoop, 2018). It may also be linked to a sense of place. In Wales, opposition to the Gwynt y Môr offshore wind project was related to the attachment that residents had to the town of Llandudno, known for its Victorian heritage and scenic beauty. The more attached a resident was to the town, the more they opposed the wind farm (Walker et al, 2011).

Our experiences of a landscape are based on what we see and hear, what we can reach out to and touch. It might come from the silence of a long walk or the joy of hearing birdsong. It might be the setting of the sun on a distant horizon or the glistening of the waves as the tide comes in. Wind turbines change these connections, altering people's sensory experiences of an area. The horizon may now be dotted with tall towers with spinning tops. A walk across rural uplands may now take in tall monuments, fenced off and badged with the logos of distant companies. This can also occur through the blinking lights placed on turbines to ward off aircraft or the flicker of shadow and sound generated by turning blades. Much of the low-frequency infrasound from the turbines is broadly understood to be outside of the average person's hearing threshold and not to pose any health risks. Despite this, the annoyance caused by the hum of wind turbines has been a complaint of many residents in areas affected (Zeller, Jr, 2010; Klæboe and Sundfør, 2016). At one wind farm in the Netherlands, 23 per cent of survey respondents reported annoyance at the sound of nearby wind turbines when they were outside (Bakker et al, 2012). Some even reported hearing it from inside their homes.

The impacts of wind power represent an intervention in which experiences of a landscape are transformed (Kempton et al, 2005; Firestone et al, 2009; Kim and Chung, 2019). People's opposition to wind farms based on a love of the landscape can overlap with other priorities, such as those linked to financial benefits or access to energy (Russell and Firestone, 2021). This can be seen in complaints about impacts on local house prices (Vyn, 2019). The average-sized wind farm might reduce the prices paid for properties within a 2km radius and which have views of the turbines by up to 6 per cent (Gibbons, 2014). This reduces as a property is further away and a turbine becomes less visible. However, while much research has been conducted on the topic, few researchers have found conclusive evidence that this is a long-term impact, with price decreases associated with wind turbines reversing over time in some places (Atkinson-Palombo and Hoen, 2014; Firestone et al, 2015).

Concerns about a changing landscape can lead opposition movements being dismissed as a NIMBY outcry of residents not wanting turbines to spoil *their* view. A telling example here is the 16-year opposition campaign against the proposed Cape Wind project off Cape Cod, Massachusetts. The offshore scheme was planned by the private company Cape Wind Associates and would have been the first offshore wind project in the US, installing 130 turbines off Nantucket Sound. Opponents to this scheme expressed concerns about marine impacts, aesthetic change, and disruption to recreational uses of the landscape (such as fishing and boating) (Firestone et al, 2009). This was a sustained, well-funded and well-connected coalition that staged a very public campaign across the media and within local, state, and federal approval processes. It was not an insular operation. Its fundraising ability was expansive. In 2003, the Alliance to Protect Nantucket Sound, a central organisation in the movement, received donations of US$1.7 million (£1.3 million), from 2,891 individuals (Kempton et al, 2005). By 2017, it was estimated that the opposition network might have raised and spent US$40 million (£32 million) (Seelye, 2017). Although the Cape Wind project did make some progress, it later struggled to raise the necessary finances and was shelved in 2018.

From the outside, there are many things that are troubling about the opposition to the Cape Wind scheme. Of the money raised in 2003, US$1.3 million (£1 million) was donated by just 56 people; four individuals gave over US$100,000 (£81,000) each (Kempton et al, 2005). Wealthy opponents of Cape Wind included the journalist Walter Cronkite, United States Senator Ted Kennedy, and William I. Koch, a billionaire who made his money in fossil fuels. Wealthy supporters helped bankroll litigation that included numerous lawsuits to block construction, and which were described by one judge as a 'vexatious abuse of the democratic process' (Trabish, 2014). Senator Ted Kennedy was linked to numerous attempts to scuttle the project which was due to disrupt his sailing routes and views from the Kennedy family's compound in Hyannis Port (Schoetz, 2007). His nephew, Robert Kennedy, Jr,[2] wrote in the *New York Times* in 2005:

As an environmentalist, I support wind power, including wind power on the high seas. I am also involved in siting wind farms in appropriate landscapes, of which there are many. But I do believe that some places should be off limits to any sort of industrial development. I wouldn't build a wind farm in Yosemite National Park. Nor would I build one on Nantucket Sound.

The opposition to Cape Wind has several lessons for a just energy transition. First, a small group can have an outsized influence on the wind project, elevating their concerns over the importance of decarbonisation agendas.

This highlights an example of how the rich and powerful can dictate the forms that renewable energy transitions take – and where. Second, this opposition emphasises how these groups can also come to dominate how a host community's response to wind energy infrastructure is presented and understood. The rich, powerful, and famous opposing wind farms often gain attention that outweighs and overshadows the other critics of such schemes. This can be seen in the UK, where national media has reported on the Rampion 2 project being questioned by the media personality Amanda Holden (Taher and Hind, 2022). Elsewhere, celebrity opposition to wind farms off the coast of Suffolk, including Dame Joanna Lumley, has received national coverage (BBC, 2022b). This media attention can lead to the concerns voiced by the rich and famous being elevated over the anxieties or priorities that might encourage others in the community to oppose or support a project.

Wealthy elites can challenge or alter energy transitions or seek to exempt themselves from their impacts (Jackson, 2010). It is difficult to deny that this is partly the case at Cape Cod, particularly when reading Walker Cronkite, the famous broadcast journalist and local resident, quoted in the *New York Times*: 'The problem really is NIMBYism ... and it bothers me a great deal that I find myself in this position. I'm all for these factories, but there must be areas that are far less valuable than this place is' (Burkett, 2003). Yet, the opposition movement to Cape Wind was more than just Walter Cronkite and the Kennedys. At the time of Cape Wind, one survey found that only 25 per cent of Cape Cod residents supported a large offshore windfarm – compared to 78 per cent of residents in Delaware, further down the Atlantic coast (Firestone et al, 2009).[3] Opposition groups included business owners, the local fishing industry, and Indigenous American Peoples. While the concerns of the Kennedys were primarily visual, others cited place-based impacts including disruption to maritime navigation, threats to habitat, and damage to spiritual heritage (Goodnough, 2010; Seelye, 2017).[4] Some residents maintained that the entrance of the Cape Wind project represented the enclosure of what was a marine commons enjoyed by all.[5] For many, the promise of low electricity rates did not outweigh the potential impacts (Firestone et al, 2009). These voices were often overshadowed by the powerful and famous, leading to the opposition to Cape Wind becoming characterised as primarily one of wealthy residents seeking to protect their landscape.

A wind project, be it offshore or onshore, is not just a project of engineering. Instead, its planning and construction are intricate social and political processes in which residents, project developers and local and national governments interact and speak to one another. Local opposition to wind power cannot be distilled down to one or two causes. It is complex and formed of overlapping concerns about fairness in decision-making.

Rural communities at the energy periphery can face large energy projects being applied by central planners and push back to call for more influence in decision-making. As Audra Parker, a prominent member of the opposition to Cape Wind has explained: 'We support renewable energy, assuming it's responsibly sited, cost-effective, and has the support of the local community. Cape Wind had none of those. Involving the community early on is extremely important' (Motavalli, 2021). Patterns of support and opposition are often uncertain and dependent upon more variables than spoilt views. Even in Cape Cod, support for offshore wind increased when survey respondents were told that the wind farm would be the first project in a broader large-scale deployment across the region (Firestone et al, 2009). Casting host communities in a negative light of NIMBYism exclusively overlooks responses linked to procedural justice, of whose voices are heard and listened to and when. There is something else at play here, linked to trust and accountable decision-making.

Social acceptance and procedural justice

Across the project approval process, numerous viewpoints are presented, impacts raised and promises pledged. Due to the role of local opposition in creating a bottleneck of new wind energy projects, many have sought to understand this process by detailing the social acceptance of energy infrastructure, both in terms of what acceptance might look like and what factors might influence levels of acceptance. As Wüstenhagen et al (2007) have detailed, this acceptance comes in different forms:

- **Socio-political acceptance:** based on the acceptance of a particular technology by members of the public and other groups (such as policy makers). For example, do policy makers accept a wind turbine as a legitimate renewable energy technology?
- **Local acceptance:** based on the acceptance of a particular project at a certain location by the local population and government. Does the local community accept the wind turbine being sited *here*?
- **Market acceptance:** based on the acceptance of technological projects and services by investors and the public as consumers as viable and appealing. Do consumers accept installing rooftop solar as a viable, cost-effective way to generate electricity for them to use?

While these forms of support overlap, energy projects being built in particular areas and landscapes require 'local acceptance'. This is influenced by numerous elements: including age, broader political beliefs, knowledge of renewable energy, length of time spent living in the area, or environmental attitudes (Ladenburg and Dubgaard, 2007; Bidwell, 2013; Ellis and Ferraro,

2016; Cashmore et al, 2019). Many have assumed that local acceptance can be encouraged by information campaigns and consulting with the host community (Wolsink, 2007). Consultations often accompany a project from planning to approval. The consultation for Rampion 2 is taking place at a formative stage of the process, before project designs and impact assessments are formalised. Rampion's consultation efforts, taking place over nine weeks (rather than the four-week period that is required), include consultation leaflets dropped through letterboxes, targeted email newsletters, local information points and direct engagement with community groups.

Procedural justice represents both a cause of and explanation for distributive injustice, as well as a potential route to address such inequities.[6] One of the primary reasons for the costs and benefits of projects and policies being unfairly distributed in an energy transition is that the decisions that drive them are made by people who benefit from change but do not bear any costs. In renewable energy transitions, examples of procedural justice are evident in the sharing of information between the developers and other groups, the opportunity for residents, communities, and institutions to participate in decision-making and influence subsequent outcomes, and a good relationship between developers and other groups (Walker and Baxter, 2017a). The absence of such conditions can motivate opposition to energy projects (Wolsink, 2007; Rand and Hoen, 2017). A set of core principles for procedural justice might include:[7]

- **Fairness in procedure:** sustained discussion over time, rather than sporadic consultations with set time limits.
- **Fairness in design:** decision-making procedures that are designed to eliminate inequalities and power imbalances.
- **Inclusiveness:** equal recognition that all should be involved in decision-making processes and that extra effort be made to include those whose participation might be restricted by other factors (such as money or time). Meetings need to be accessible to all and not just in terms of physical access – they should be timed so that as many residents as possible can attend.
- **Shared information:** the provision of accurate information through appropriate channels, allowing all participants to be involved in the decision-making process on a relatively equal footing.
- **Shared authority:** no party has the ultimate power to overrule the concerns or recommendations of others.
- **Authoritative decision-making:** the recommendations of participants are final, rather than seen as something that can be ignored.

Procedural justice is built on trust, with host communities being listened to and engaged with in good faith (Dwyer and Bidwell, 2019). Members of the public are often more likely to accept a wind energy project if they are able to participate in the planning and decision-making process and influence outcomes or if they perceive greater benefits from the project itself (Liebe et al, 2017; Walker and Baxter, 2017a; Mills et al, 2019). However, such expectations can remain unmet.

While planning processes, such as those in the UK, provide safeguards for procedural justice, the influence that host communities have on a project may be limited. Both Rampion 1 and 2 are Nationally Significant Infrastructure projects, taking ultimate decision-making power away from local authorities and gifting it to the national government. During its approval process, Rampion 1 was subject to concerns about visual impacts on the landscape, particularly to the famous views from the South Downs National Park. However, the Examining Authority, appointed to chair the process, concluded that: 'although the visual effects of offshore development upon the National Park and Heritage Coast cannot be eliminated, the level of benefits to be afforded from the proposed wind farm in terms of the need for energy infrastructure ... outweigh the level of damage likely to be occasioned' (Walker et al, 2014: 123). Across the Examining Authority's report, the national need for renewable energy outweighed landscape-based impacts and local economic concerns (Rydin et al, 2015). The Rampion 1 consultation was criticised for not doing enough to directly reach residents (such as those who might not have access to email) and for not actually listening to concerns voiced. As one local fishing business owner explained to researchers: 'I do feel that we had an opportunity to say everything that we wanted to say but I don't necessarily feel like it was heard all the time' (Rydin et al, 2018: 278).

Energy transitions are often directed by energy authorities, regional or national governments, or private developers. The story of energy projects is often one of a national government approving a project that will be built by a private company, with the residents unable to influence proceedings and promises of economic benefits unclear. The concerns of host communities are neglected, cast aside, or characterised as 'NIMBYism' (Ottinger, 2013). As such, opposition to wind projects can often be a push-back against top-down approaches to energy transitions, protesting decision-making processes rather than only resisting the project itself. To overcome such tensions, project developers often provide community benefits and payments as compensation for impacts and to convince residents that a new project would benefit them.

Compensation and windfalls

To address the concerns of residents and go some way to lessening the distributive injustices associated with wind farms, many project developers and energy companies dedicate funds to community work, outreach and benefits (Lienhop, 2018; van Wijk et al, 2021). On the Sussex coast, Rampion dedicated £1.7 million to 153 local projects between 2017 and 2021 (Rampion Offshore Wind, 2022). Work funded has included providing sensory plants and raised beds at Hove Lagoon, a new accessible jetty for the Littlehampton Sea Cadets, and support for young carers in Brighton and lifeguards in Seaford. Community benefit schemes can also seek to provide opportunities for those living in the area. At the site of Project Fortress in Kent, the site discussed at the beginning of Chapter 2, local communities previously received a £850,000 benefits package from the developer of the London Array, another nearby offshore wind farm whose connection reaches land nearby (Cowell et al, 2012). This included bursaries for local students to go to university, donations to local schools, and funding for nature conservation (Cowell et al, 2012).

Compensation and community benefit programmes like this are rarely standardised. The forms they take differ across developers and respective projects. A project developer or owner often sets aside an amount of money to finance projects in the community. This amount may be funded through different ways, such as a) an agreed annual sum paid per megawatt (MW) of installed capacity, b) a variable amount linked to total electricity generation or profit, c) an agreed lump-sum payment, or d) a mix of all of these. Specialist schemes might be provided to support the community in a defined way. The renewable energy company, Thrive Renewables[8] provides a community benefits programme, through a partnership with the Centre for Sustainable Energy to fund and support energy efficiency improvements in local, community buildings such as village halls and community centres (Thrive Renewables, nd.). The aim is to improve these spaces and make them more accessible to the community, by reducing running costs and enabling them to stay open to the community for later and for longer throughout the year.[9] This scheme is not based on the distribution of an agreed percentage of revenues. A pot of money is made available to communities, who can request funds (of up to £4,000 per grant) for energy efficiency projects. The fund involves assistance in gauging the energy footprint of buildings and how it can be lessened. This has allowed the funding of simple cost-effective measures, from LED lighting to heat-pumps.

With wind farms often sited on the energy periphery and in regions of below-average incomes, the financial benefits of these schemes can be transformative (Cowell et al, 2012). Households, communities, and local organisations might find new sources of income that can, in turn, be reinvested in the community. It can also go some way to lessening opposition (Brennan and van Rensburg, 2016). In a choice experiment run with rural residents in Germany, 60 per cent of participants voiced support for local wind energy if they received financial payments – in this case, in the form of reduced electricity bills (Lienhop, 2018). This provision of community benefits schemes also represents a form of *justice as recognition*, with project developers recognising that a wind farm is being built on a landscape of importance to the host community and that any disruption or impacts should be compensated (Rudolph et al, 2018).

While residents might see financial benefits and payments in different ways, their fair distribution is an important factor in determining local support, with many preferring public programmes that include many people and communities, rather than individual compensation arrangements (García et al, 2016; Walker and Baxter, 2017b). Residents who are suspicious of project developers can see community benefits packages as a bribe or as schemes that merely compensate for impacts rather than attempting to change attitudes toward the project (Aitken, 2010; Cowell et al, 2012). A challenge of procedural justice in this sense can be found in pinning down who we mean by 'community'. Who will benefit or be impacted, and who won't? The impacts of wind energy flow across landscapes, meaning that those residents and communities who perceived themselves as 'impacted' might be geographically spread out and live a great distance from one another (van Wijk et al, 2021). In their consultation, the team at Rampion 2 engaged with residents across an expansive area – along both the transmission cable route to the proposed substation and the coastline from Selsey Bill in the west to Beachy Head in the east, as well as those living along the east coast of the nearby Isle of Wight (Rampion 2, 2021). These are large areas to be consulted and compensated, including coastal, urban, and rural areas that are collectively home to over one million people.[10]

At some wind farms, compensation schemes are on an individual basis, with money given directly to those who own the land where turbines are to be raised. This might be in the form of a one-off payment or an annually paid rent. In the United States, it has been estimated that landowners receive close to US$222 million (£183 million) per year in such payments from wind energy developers (Adelman, 2020). These payments, often between US$3 and US$7,000 (£2,500–£5,700) can bring important benefits. They might provide new income to farmers struggling financially in return for what is, in reality, a small parcel of land. This money may be reinvested in the local economy, be it through buying goods or developing new community initiatives (Copena and Simon, 2018; Adelman, 2020).

Patterns of land ownership can also highlight an injustice of wind energy compensation, with many who own land and receive compensation not necessarily living in the community itself nor spending money in the local economy. For example: in Michigan, the United States, an estimated 23.8 per cent of compensation payments to landowners affected by the Garden Wind Farm went to absentee landlords in 2016 (Adelman, 2020).

Compensation payments can reward inequitable patterns of land ownership and the monopolisation of land by the rich and powerful in the region.

In the UK, a small number of landowners have gained financially from the expansion of onshore wind power, due to a broader inequality of who owns the land where these projects are built. Half of the land in England is owned by less than 1 per cent of its population (Shrubsole, 2020). In 2014, it was estimated that only 432 private landowners owned 50 per cent of all private land in rural Scotland (SLRRG, 2014). The amounts that some individuals have received from past energy projects are staggering. In 2012, it was reported that the Earl of Moray had received up to £2 million per year in rent from the 49-turbine project built on his land in Perthshire, Scotland (Vidal, 2012).

The rewarding of some with compensation, while excluding other residents, can, in turn, create tensions within the community (Tolnov Clausen and Rudolph, 2019). Some might be compensated, while their neighbours put up with the impacts without redress. In Argyll and Bute, Scotland, the expansion of onshore wind schemes was accompanied by the ad hoc growth of numerous community benefit schemes, different from developer to developer, and directed towards the community council areas where turbines were directly located. The result was that some communities with turbines sited on their doorsteps saw investment in pipe bands and golf courses and display boards, but those living elsewhere in the affected landscape did not receive any compensation or witness investment despite them also being impacted (Cowell et al, 2012). This, in turn, generated tensions between the 'local' and 'wider' communities in the region (Cowell et al, 2012). To remedy this, Argyll and Bute Council organised a new Community Wind Farm Trust Fund which recommended a minimum payment of £2,000 per MW of installed capacity be paid into a local trust fund. Funds are inclusive and available to the wider community, be they living on the land where turbines were built or further afield. The figures raised can be transformative over a project's lifetime project – with a £50,000 annual community payment from a project developer equalling an investment worth over £1 million in the local economy over a 20-year duration of an energy scheme (Argyll and Bute Council, nd.). Yet, for community benefits to be sustained fairly, joined-up thinking across different districts and regions is required.

The income gained from offshore wind in the UK has been heavily linked to one family. Operating through the Crown Estate, which manages a broad property portfolio, the Royal Family owns the seabed that extends around the UK.[11] The 2021 auction of rights to the seabed (the first such auction in a decade) by the Crown Estate attracted frantic bidding from energy companies, ultimately leasing rights for six new offshore wind farms. Funds raised this way are, at first, directed to the UK Treasury, which keeps 75 per cent of the amount raised but returns 25 per cent to the Crown in the form of its annual Sovereign Grant. In 2023, the Sovereign grant, the rest of which is funded by tax-payers, equalled £86.3 million a year. The further auctioning of the seabed for offshore wind farms might increase this substantially – the 2021 auction alone generated a total annual windfall of up to £879 million, of which up to £220 million might have been returned to the Crown every year. In January 2023, the Crown Estate confirmed agreements for six new offshore wind projects, with it reported that King Charles III had asked for the profit of this windfall (estimated to be close to £1 billion a year) to be redirected towards the wider public good. Such a move was celebrated as a diversion of profits to help those struggling in a cost-of-living crisis. However, it did not address the fact that the funds generated from renewable infrastructure continue to flow *through* the Crown Estate, while highlighting that they can choose to forgo these revenues. This process is subject to Treasury-set rules but, unless these rules change, wind energy projects might continue to generate ever-larger revenues for the Royal Family in the future.

From planning to generation, wind turbines are socially embedded and contextually dependent, raising questions of who owns the land where energy projects are built – and who owns (and benefits from) the electricity generated. Wealthy landowners have looked to booms in onshore wind power as a source of profit. With offshore wind in the UK to be expanded and onshore wind likely to return, it is important to reflect on who might financially benefit from these projects. In an era of rising energy prices and increased household and national energy insecurity, this raises important questions for a just energy transition. Who is generating energy and to benefit whom?

Who owns the wind?

The value of a wind farm, including its costs, benefits, or the balance between them – is more than economic. Who owns a wind farm has an important influence on the acceptance of wind projects by host communities. Opposition to wind farms can often be directed at project developers and how the scheme is planned and approved, particularly if communities are treated as passive bystanders to be consulted, informed, and compensated (Bidwell, 2013). A wind farm becomes a top-down intervention in the landscape by external groups for the benefit of those living a great distance away. The list of companies who were successful in the 2021 and 2022 auctions of offshore rights by the Crown Estate in the UK included TotalEnergies, BP, Shell, and Iberdrola, all expansive multinational energy companies (Crown Estate, 2021; Crown Estate Scotland, 2022). The renewable energy landscapes created become inscribed with both a distributive and procedural injustice. While host communities will experience changed horizons, the profits will be enjoyed by shareholders and executives who may never set foot on the shorelines affected.

On the Isle of Lewis in the Scottish Hebrides, the building of a new wind farm near Stornoway created tensions, not around what would be built but over who would own it. EDF Energy had planned 36 wind turbines to be built on parcels of moorland in the area. The land was legally held as a commons and local crofters had access to this landscape to graze animals and source resources, such as peat. Despite this common use of the land, the ten elected trustees of Stornoway Trust granted a 70-year lease to the land for the wind farm. Over 200 crofters lodged objections and, as an alternative, proposed a community-owned wind project that would invest profits in the area. There is precedent for such a scheme. In nearby Point and Sandwick, three community-built turbines provide revenues that support an arts centre, a drug and alcohol addiction programme, and a hospice in the local area (Watt, 2018). In 2020, Scottish judges found against the crofters, ruling in favour of EDF and the lease granted by the Stornoway Trust as the owner of the land. The grounds for the Court's decision were that any approval of the crofters' case would be detrimental to the landlord This ruled in favour of the entrance of a large international company into the landscape, with little recompense beyond a promise of a community benefit scheme.

Wind energy projects in the UK primarily remain in the hands of private companies, international investors, and pension funds. The London Array is jointly owned by RWE AG (the German multinational), Ørsted (the Danish partly state-owned renewables company),Masdar (Abu Dhabi's renewable energy company) and the Canadian *Caisse de Dépôt et Placement du Québec* pension fund (London Array, nd.). Whitelee Windfarm, the UK's largest onshore project, found on Eaglesham Moor across East Ayrshire and South Lanarkshire is ultimately owned by Spanish firm Iberdrola (through its subsidiary Scottish Power).

The Rampion projects are owned by RWE AG, Enbridge (an energy company from the United States), and the Offshore Wind Consortium formed of long-term investors (Rampion, nd.). I didn't know this while I moved beneath the turbines offshore, but the Universities Superannuation Scheme – my own pension fund and the UK's largest private scheme – is part of this consortium that owns 25 per cent of Rampion. The electricity generated off the coast of Sussex is, in theory at least, providing me with my own retirement income.

Offshore wind is a terrain of financial investment and returns – of long-term investors, focused on protecting investments and driving up returns, paying out dividends to shareholders and increasing retirement incomes. New wind farms in the UK will be built by BP, Total and Shell on land sold by the Crown Estate. Despite promises of compensation payments and benefit schemes, the funds provided by offshore wind farms can be meagre. In Scotland, such funds pay out just £150,000 between them (Dobson, 2022). Collectively, the communities living near onshore wind projects receive £22 million a year – which is no small sum but is 0.6 per cent of the value of the electricity produced by the infrastructure outside their homes (Dobson and Matijevic, 2022).

In a just energy transition, wind energy should support and enfranchise the communities in places where infrastructure is built. They must be given a stake, a sense of ownership and a connection that allows the perception that the wind farms belong on the landscape. One example of such locally owned renewable energy can be found on the island of Samsø. Found in the Kattegat Straight, between Jutland and Zealand in Denmark, this island is home to less than 3,800 people. In 1997, the Ministry of Energy in Denmark launched a competition for islands and regions to submit plans to achieve 100 per cent self-sufficiency and transition to renewables in 10 years. Samsø, working with communities on the Thyholm peninsula and three other islands, was announced as the winning project in October that year. The competition, part of Danish national decarbonisation policy, was devised to identify a new, model renewable energy community – that could both show such a transition is possible and provide insights into good practice (Mundaca et al, 2018). Since 1997, Samsø has transformed its energy system, achieving 100 per cent renewable energy generation in 2005 and becoming a poster child for energy transitions. Energy is provided by solar and wind generation, with surplus wind energy exported to the mainland, and warmth provided by district heating plants (Marczinkowski et al, 2019). Profits from its offshore wind farms fund the Samsø Energy Academy, which has welcomed journalists, researchers, and communities from around the globe to learn from the island's experience.

The community of Samsø owns the wind. Not private developers or companies based in different towns, cities, or countries. This has enabled communities and residents to not only be part of energy transitions but to define it. Farmers clubbed together to build turbines on their land, taking out loans from banks to pay for it (Connelly, 2021). Locals established their own district heating plant in the village of Onsbjerg. The municipal government bought five offshore wind turbines. Two others are owned by local energy cooperatives. Both are accountable to the community (through direct elections and member control respective) and use the revenues to support further sustainability work across the region. This ownership, centred on the community, has helped generate support, buy-in and a connection across the community, with many residents generating their own energy. Community organisations, companies and volunteers have played a central role in the energy transition on Samsø, influencing plans and benefiting from consultancy on potential energy savings.[12]

Internally, Samsø's energy transition is grounded in local community identity, which has provided a powerful motivation for action, with many residents holding pride in the island's role as an exemplar in energy transition (Mundaca et al, 2018). Cooperative management of resources is rooted in the island's history. Samsø has historically been a rural, agricultural economy and the community living there have experience in the shared management of dairies and a vegetable packing and distribution facility (Sperling, 2017). For many, farming culture and tradition remain an important part of the island's life – leading to a centrality of trust, shared decision-making, and collaboration as key principles that underpin Samsø's energy transition. This context also highlights how community energy can stem from a time of crisis. Winning the national competition in 1997 coincided with the closure of the local abattoir on Samsø, the biggest employer on the island in 1997. While unrelated, this may have provided a sense of togetherness across the island community, detached from the mainland, in a time of potential economic crisis (Sperling, 2017). As Samsø's Mayor has previously told reporters: 'Society was in crisis … Unemployment was high. It could be 16 per cent in winter, maybe 7 per cent, 8 per cent in the summer. We had lots of depopulation and people wondering, well, how can we create a future for the island?' (Connelly, 2021). The 1997 award came at a time Samsø was changing and the advent of renewable energy on the island became a vehicle for how the island might improve and find a new future.[13]

Rule 2: Elevate and emphasise the participation and voices of communities when developing new energy projects

The wind is ultimately public. It flows through land and property, paying no attention to ownership and tenure (Hughes, 2021). If wind turbines come to represent procedural injustices, they might also be put to use to provide a way for a community to have more of a stake in energy transitions. Community acceptance and 'buy-in' into a project can be encouraged through changing ownership of the infrastructure itself. Wind energy has always been linked to a sense of place. In the UK, some of the first efforts to harness the wind for energy negotiated different conceptions of how it was linked to the landscape and certain places, and the relationships between national governments and local communities (Dudley, 2020; 2021). This continues, with wind turbines both becoming inscribed by preexisting forms of place attachment and creating new ones.

Those living close to wind turbines have been found to adjust to the aesthetic changes over time, even becoming proponents for wind energy projects (Meyerhoff, 2013).While for some wind turbines might represent a blight, for others, they can come to represent something more positive. In Orkney, one resident organised for the wall in their front garden to be lowered – so that they would have a better view of the turbine blades to count how much money they earned with each rotation.[14] Elsewhere on the Orkney Isles, the Rousay, Egilsay, and Wyre Development Trust (active across three northern islands of the archipelago) has decorated its wind turbines with a local artist painting around the hands of local residents, symbolising a connection between residents and technology. On the Isle of Gigha, in the Hebrides, residents have dubbed locally owned wind turbines as 'the Three Dancing Ladies' (Warren and McFadyen, 2010). These turbines, part of the first grid-connected community-owned project in Scotland have previously, raised annual profits of £85,000 – far more than the community benefit payments made by private-owned wind farms nearby (Warren and McFadyen, 2010).

A just energy transition must seek to treat local communities as more than just the hosts of new infrastructure. Instead, a transition must seek to redefine the relationship between community and infrastructure. This might involve a degree of co-ownership of a project or sitting on decision-making bodies or gaining income from its generation in the form of dividends (Lienhoop, 2018). Once the wind and the infrastructure that catches it are public-owned and controlled, its energy can be used in different ways and for new purposes . Those living near turbines might be given access to cheaper tariffs. In the UK in 2022, Octopus Energy created a new tariff for customers in Caerphilly and Market Weighton, towns local to the company's two biggest wind turbines, with residents able to enjoy locally generated renewable energy. Funds from community benefit schemes from large renewable energy projects can also be directly channelled into locally owned renewables, providing for local energy independence and security.

Wind projects owned by local authorities or cooperatives have been found to gain higher levels of local acceptance (Haggett, 2011; Musall and Kuik, 2011). On Cape Cod, where extensive opposition stopped the Cape Wind project, 22 per cent of survey respondents were more likely to voice support for offshore wind if it was to be built by the local government rather than a private developer (Firestone and Kempton, 2007). It is often easy to see why. Incomes gained through local ownership are often far more than what is paid by a developer's community benefits scheme. Community-owned wind farms in Scotland have provided, on average, 34 times more financial benefits than those owned by private companies (Aquatera, 2021). Revenues from the decorated turbine owned by Rousay, Egilsay, and Wyre Development Trust fund numerous projects and grants, including support for local internet installations, bursaries for driving lessons, and providing books of ferry tickets to local children (REW DT, nd.).[15] All investments that can be transformative for those living in these communities.

Renewable energy projects can fulfil the tenets of procedural justice when they empower and support communities. Local authorities and organisations retain an important role in doing this, ensuring that communities are both protected and empowered within new energy schemes. There are myriad examples of local authorities and communities taking an active role in planning, building, and managing wind energy – and other forms of renewable energy too. Rather than schemes approved by central governments, built by private developers, and operated by shareholder owned energy companies; these efforts are decentralised and community-focused. Ventures can come to represent a just energy transition – in which wind power is put to use by the community, and for the community's benefit.

5

Community

St Bede's Catholic College in Lawrence Weston sits at the northern edge of the city of Bristol in the southwest of England. Behind it lies flood plain and the roar of the motorway. In the distance is Avonmouth, the industrial heartbeat of Bristol. Decades ago, schoolchildren would look out of classroom windows and see smoke of different colours rising from furnaces and smokestacks. Greens, whites, oranges, and blues. The flow of the smoke could be read to show the strength of the wind that day. On other days, visitors might come to the school to take blood samples from the students, for fear of the lead emitted by the petrol of the cars journeying nearby.[1]

The school is different now. So is the view from the playground. Some things have stayed the same. The gas-fired Seabank Power Station still stands on Hallen Marsh to the north. Avonmouth Dock remains an expansive hub, connecting Bristol to global trade. The area also remains home to many industrial sites. I live nearby and we get a booklet dropped through our doors every year, advising what to do if we hear the 'Severnside Siren': an alarm, named after the nearby river, that warns us if a dangerous substance is accidentally released. However, the landscape is different; gone are many of the chimneys and the smoke acting as a weathervane. In their place lie wind turbines – some on the grounds of the nearby wastewater treatment plant owned by Wessex Water and Thrive Renewables, others at the Avonmouth Dock estate, one at a nearby wine-bottling plant, and more at the site of a former Shell oil tank.

A new turbine will soon join them. The previous construction of many wind turbines on this landscape was seen by many as neglecting the local community, with projects built outside residents' windows but not always providing opportunities for their participation, engagement, or benefit.[2] Soon one will be the community's own. Initially stemming from an idea between residents in a pub, the project is no small feat. It will be the tallest onshore turbine in the country and will generate enough electricity to power 3,000 homes. It will be community-owned and community-led via Ambition Community Energy, an offshoot of Ambition Lawrence Weston, a local organisation active in the neighbourhood. The original idea of the Lawrence Weston turbine was to directly supply cheap electricity to residents, but the cost of transmission lines made this plan too difficult. Instead, electricity generated will be sold to the national grid and the revenues invested in the local community.[3]

Ambition Lawrence Weston is an integral organisation in the area. It has worked to save a youth centre, supported 40 local organisations and numerous community schemes, lobbied to bring a supermarket to the neighbourhood, and replaced missing bus routes to help residents get to work (ALW, 2022). Revenues from this turbine will further support this important work: perhaps being used to support residents in paying for energy efficiency measures. Or, with many dependent on prepayment meters, it could be used to support them in paying their energy bills and keeping the lights on. Near the turbine, a learning zone will be set up to help local children, like those at St. Bede's, to learn about renewable energy. A live data feed will be set up so that schools can see how much electricity is being generated at that exact moment.

The scheme in Lawrence Weston highlights how community-centred energy projects can create new opportunities for groups and organisations to engage with one another and influence local or regional politics or economies. There are many possibilities, both for energy transitions and for how renewables can become part of broader community wealth building to address economic problems. Wind turbines could power local homes and village halls. Or like at Lawrence Weston, electricity could be sold and the revenues returned to the community and neighbourhood.

This book calls for a just energy transition formed of a nationally led but community-centred energy model. Such a model requires an examination (and reformulation) of the relationship between the national government, local and regional authorities, and communities themselves. Community energy schemes are rarely small and easy to start. Fundraising and technical support are required to get these schemes out of the imaginations of those who dream them up. Community-centred projects require commitment, financial backing, and sustained work. The onshore wind turbine at Lawrence Weston cost more than £4 million – no small amount. Its planning and construction was supported through financial means and technical support provided by regional authorities (such as the West of England Combined Authority and Bristol City Council), private companies (Thrive Renewables), local organisations (Bristol Energy Network), sectoral groups (Community Power Solutions), charities (with 'Big Local' funding from the National Lottery), and private benefactors. The onshore turbine is found on land owned by Bristol City Council and received funding from myriad public and private sources. Andrew Garrard, a pioneer of modern wind engineering and Bristol resident, helped designed the turbine. In pursuit of this project, this coalition overcame numerous hurdles: including high costs, a planning process that leaves little room for onshore wind, no financial support for community energy from the national government, and challenges in the planning process.[4]

In this chapter, I explore these possibilities and focus on the policies, mechanisms and opportunities that might support community energy in a just energy transition, and how such elements can define the forms that decarbonisation takes. For community-centred energy projects to move beyond being local, small-scale endeavours they need supportive and active policy from central governments. Such support can catalyse community-centred energy schemes through the building of direct links between energy projects and local wealth generation. The first step towards doing so is the recognition of the significant role that local groups can play in achieving both.

Energy transitions and misrecognition

The past twenty years have seen the rapid expansion of interest in community renewables, with researchers, policy makers, and communities themselves exploring how local groups can take greater ownership and control over energy transitions. Community-owned renewables can, in theory, allow for energy transitions to move beyond the problems of distributive and procedural justice highlighted in previous chapters. In doing so, they can also provide an important route of *justice as recognition* through the acknowledgement and understanding of the role that communities can take in decarbonisation, and how some communities need more support than others. This would involve an understanding of social justice that includes both the redistribution of benefits and impacts and the recognition of different hopes, values, experiences, and outcomes (Fraser, 1999).

Justice as recognition in energy transitions has been broadly understood as the need to recognise the roles of different groups and communities in energy transitions and ensure their representation in policy (Sovacool and Dworkin, 2015; Knox et al, 2022). An approach of justice as recognition is based on the need for groups, institutions, policies, and people to acknowledge or appreciate the differences in the various experiences, motivations, priorities, or actions of others (Whyte, 2018). Injustice is rooted in the failure to take differences between and within communities, households, and people into account when making decisions. This can be linked to the neglect, disrespect, or exclusion of others in which people and communities are seen as 'less than' others (Fraser, 2000). Such a 'misrecognition' can take place in several ways: belief or value systems held by some might be deemed unworthy or overruled by dominant ways of seeing things, groups might be subject to persistent, entrenched stereotypes or their views and aspirations belittled, or remain entirely unacknowledged and pushed to the edges of politics or policy.

In all cases, people, groups and communities are unable to meaningfully participate in society or have their voices, aspirations, or anxieties heard in decision-making (Fraser, 1999; 2000).

Misrecognition often overlaps with processes of stigmatisation, in which certain groups, communities, or neighbourhoods are condemned as a 'problem' in society or deemed in need of certain interventions but not others. This is evident in narratives around the austerity politics of successive Conservative governments from 2010 onwards in the UK, who argued that they could reduce budget deficits through the stripping of public services and reduction of public spending. These policies were often accompanied by assertions that poverty was linked to behaviour and personal choices, with those claiming state welfare benefits deemed as lazy, living off the hard work of others, or in disadvantaged circumstances due to their own poor decisions (Pemberton et al, 2016). These narratives became commonplace and reshaped popular understandings of what poverty was and how it was experienced (Shildrick, 2018). Such a characterisation led to a series of punitive measures for those on low incomes or out of work such as the withdrawal of financial support. This narrative of stigmatisation and misrecognition has continued today. The Home Secretary, Suella Braverman argued in October 2022 that: 'There are too many people in this country who are of working age, who are of good health, and who are choosing to rely on benefits, on taxpayers' money, on your money, my money, to get by' (Daisley, 2022).

These characterisations of those living in poverty represent a process of 'misrecognition', in which those claiming state support were demonised but the complex reasons behind their circumstances were not understood or given space in policy. Broader societal causes of poverty and disadvantage, fragile personal and family financial situations, and the proliferation of low-paid, precarious work and zero-hours contracts in the wake of an economic recession remained unacknowledged (Pemberton et al, 2016).

A narrative of 'choosing' to live in poverty can also be found in the stigmatisation and misrepresentation of those experiencing energy poverty, who are unable to access the amount of energy that they require, often due to high costs. They have been presented by some as being responsible for their circumstances due to their energy usage (Simcock et al, 2021). Previous policies have adopted a similar approach, presenting problems of energy poverty as the result of 'inefficient' use of energy and the need to 'nudge' people into 'better' uses of electricity and heating (Snell et al, 2015; Jenkins et al, 2016). Within such an understanding, energy poverty is seen, by some at least, an individual issue, with those living in it stigmatised and deemed to blame for their own circumstances (Simcock et al, 2021).

This misrecognition of energy poverty signals an important overlap between justice as recognition and the forms of distributive injustice discussed in Chapter 2. These links can be seen in how policy makers understand problems in narrow ways that overlook important causes or patterns of injustice. Policies that both overlook the causes of vulnerability to energy poverty and disrespect those suffering from it can allow the conditions that lead to energy poverty to persist and worsen. While energy poverty is primarily a problem of distributive justice, it is inscribed with a misrecognition of the different needs and vulnerabilities of those at risk (Walker and Day, 2012). Those suffering from energy poverty become invisible in policy, merely being encouraged to do better by using electricity more efficiently.

Injustices of misrecognition, distribution, and procedure overlap and underpin one another (Coolsaet and Néron, 2021). The links between the procedural justice discussed in the previous chapter with injustices of misrecognition are evident in the limited roles that local communities can to hold in decarbonisation and the siting of renewable energy projects. Patterns of misrecognition in social or economic policies can become 'institutionalised' and embedded within a community's links to wider processes and opportunities placing it at the periphery of decision-making (Fraser, 2000; Simcock et al, 2021). Certain groups might be ignored or misrepresented in energy policy-making, resulting in them being unable to meaningfully participate (Knox et al, 2022). Elsewhere, many project developers can misrepresent and stigmatise those residents questioning renewable projects as a form of NIMBYism. This assumes what is called a 'social gap', believing that most people generally support renewables, just not where they live (Jenkins et al, 2016). This is a process of *misrecognition*, in which local communities are seen as either passive recipients of change or obstacles to be overcome. This overlooks the variety of alternative routes and forms of decarbonisation that are emerging from local communities and organisations.

Despite patterns of misrecognition in energy transitions, community-centred projects and innovations are providing solutions to local energy needs (Knox et al, 2022). The wind turbine at Lawrence Weston is such a solution that pushes back against persistent processes of misrecognition. Understanding the transformative potential of the turbine is inseparable from the political and economic context of the neighbourhood itself. Parts of Lawrence Weston are among the most deprived areas in England (Bristol City Council, 2022). 27 per cent of the working-age population of the neighbourhood are claiming some form of state benefits (the Bristol average is 14.1 per cent) and 36 per cent have no educational qualifications (20 per cent) (ALW, 2022). Lawrence Weston's historical and geographical location at the periphery of Bristol has resulted in a unique history of environmental injustice, with the neighbourhood's proximity to Avonmouth exposing the community to a mix of pollutants, with little benefit. The population of 7,000 have often been at the sharp end of change. The austerity policies of Conservative governments hit the community hard, leading to a loss of services and opportunities for many (Lacey-Barnacle, 2020). This has continued, with many in the community vulnerable to price spikes and soaring bills. In 2022, 16.7 percent of those surveyed in the ward of Avonmouth and Lawrence Weston (which includes several neighbourhoods with higher average incomes) reported finding it difficult to manage financially in the cost-of-living crisis (close to twice the Bristol average of 8.7 percent) and 6.3 percent reported visiting a food bank that year (BCC, 2022).

For a community like Lawrence Weston, a locally owned wind turbine could be transformative. Residents might better understand the contextual factors that lead to energy insecurity or problems in the area, as well as better guide the revenues to be spent productively in the local economy allowing knock-on benefits for others (Lacey-Barnacle and Bird, 2018). A just energy transition is about more than fair distribution and procedure. It should also incorporate a recognition that communities are not just tools or audiences in decarbonisation but, instead, can benefit from being involved – in terms of energy, income, and in gaining new forms of authority and influence.

People power

The term 'community' in energy transitions can often be unclear. A recent review found 183 different definitions of the term in academic research on community energy alone (Bauwens et al, 2022). This is often due to the context in which different initiatives are taking place. Some people might use the label 'community energy' to discuss energy networks that are participatory and people-led. Others might do so to describe shareholder owned energy companies that focus on channelling private economic gains towards local benefits. Others still might use community synonymously with energy cooperatives, which are owned and run by shareholder-members. Across this book, I broadly understand community-centred energy to represent energy projects that are based on local action, ownership, and/or governance and which aim to provide socioeconomic and environmental benefits for a particular community. This definition draws on others, such as Walker and Devine-Wright (2008), in understanding community energy as initiated through local stakeholder reaction and focused on local benefits. Projects that fit under this definition might include: schemes where a community group hosts energy technology on sites that they own, where members of a community have some democratic control over a project or have an active role in defining the scheme itself (Braunholtz-Speight et al, 2018). Within such ventures, residents and communities move from passive to active participants in decarbonisation, exercising a degree of control and influence over energy transition, and supporting the production of electricity as well as consuming it. All suggest an approach that characterises renewable energy as a public good can be encouraged at levels closer to home. Residents' groups, charities, churches, or local government authorities – all can become important active participants in the local manifestations of a just energy transition.

Community-centred energy infrastructure is not a new pursuit. Local-led schemes have historically emerged to fill the gaps left by national governments and large energy companies. In the US, more than 900 rural electric cooperatives have provided electricity to 42 million customers across rural regions and isolated communities (Aronoff, 2017). These customers were left out of the transmission networks built in the mid-20th century. Many electricity providers, faced with the high costs of building the transmission lines, left these communities out of their electricity networks, leading to many lacking energy access (Aronoff, 2017). This is an experience shared with communities elsewhere. In Spain, the Cooperativa Eléctrica d'Alginet was set up in the 1930s by residents seeking to address the neglect of the village of Alginet by electrical utilities, who were uninterested in connecting the community to the electrical grid. Other communities and residents in Spain have historically owned small hydropower projects or raised funds to secure a connection to regional electricity grids (Cuesta-Fernandez et al, 2020).

Energy cooperatives have also been formed as a response to dissatisfaction with incumbent electrical companies – and to challenge them. Cooperatives – such as Som Energia and Goiener in Spain have attracted tens of thousands of new customers (Cuesta-Fernandez et al, 2020). Some cooperatives have grown to a size where they are competitors to bigger energy companies. Som Energia is one of the largest energy cooperatives in the world – counting over 81,000 members, serving over 127,000 customers across Spain and producing close to 25 gigawatt hours of electricity a year (Capellán-Pérez et al, 2018). In France, Enercoop is a national network of cooperatives, working with a total of over 130 energy producers, providing electricity to 100,000 customers, and investing €98 million in local communities (Enercoop, nd; 2021). Despite being an expansive operation, Enercoop retains several core principles, including a commitment to a 'one member, one vote' form of democratic governance.[5]

In the UK, energy cooperatives exist in numerous towns and cities and landscapes. They do incredibly important work: from Brixton Energy's placing of solar panels on social housing on the Loughborough Estate in south London to Carbon Co-op's advocacy work and advice to help people reduce their household emissions in Greater Manchester. The successes of many energy cooperatives should be celebrated. They often, as in the past, plug gaps left by central government energy policy. Like the wind turbine at Lawrence

Weston, these are the success stories of energy transitions. The ground is set for there to be more stories like this, with many people interested in locally generated energy. In one survey, 79 per cent of respondents voiced interest in investing in at least one form of community-centred energy, which was preferred to investments by private energy companies (Cohen et al, 2021).

Community-centred energy projects can bring numerous benefits to the communities that host and own them. According to previous research by the UK government, community energy can provide 12 to 13 times as much social and community benefit as equivalent schemes run by private and commercial organisations (DECC UK, 2014). However, no community energy project exists in isolation. It is always embedded within local conditions that allow some to participate and exclude others. People's ability to take part in local energy schemes or the roll-out of new technologies may vary due to levels of education, available information, time constraints, or up-front financial costs (Knox et al, 2022). All suggest the need for broader approaches to ensure that a just energy transition allows the participation of as many people as possible. A just energy transition requires more than members of different communities doing it for themselves. It requires intervention from the state and national governments to devise new forms of generating and managing energy in ways that benefit local communities.

State support

Many of the examples of community-centred energy discussed in previous chapters are intricately linked to the decarbonisation policies of national governments and energy networks. The case of Samsø discussed in Chapter 4 is tied to a Danish government scheme in the 1990s that provided tax subsidies to those buying shares in wind turbine cooperatives, leading to 100,000 Danish families belonging to such ventures in 2001 (Sørensen et al, 2002).[6] In Orkney, detailed in Chapter 3, the connection to the mainland via a transmission cable has guided its energy transition in different directions, with the limits of transmission leading organisations to hydrogen technology to store electricity and use it elsewhere and at different times.[7]

Central governments influence decentralised, local and community-centred approaches, defining (consciously or otherwise) projects from the idea stage to the generation of electricity through the context in which such approaches might emerge. The expansion of community energy projects in the UK was enabled by government policy. Successive Labour governments between 1997 and 2010 launched initiatives and measures to support community action on energy, providing funds for organisations to work with and advise local communities on potential measures (Hargreaves et al, 2013).[8] The motivations behind these policies included a desire to ensure benefits for residents as a way to address opposition to wind farms, to help stimulate the emerging renewables market, and the recognition that revenues from community energy could support local areas suffering from economic decline.

It was under the coalition government of the Conservatives and Liberal Democrats (2010–15) led by the then Prime Minister David Cameron (2010–16) that community energy experienced a 'golden age' in the UK. A central element of this was the Feed-in Tariff (FiTs), launched in 2010 and which provided a guarantee of above-market rates paid for electricity generated by small-scale renewables for 20 years. FiT's have provided a dominant engine for decarbonisation across numerous countries. This is because they go some way to address a major challenge of energy transitions: namely the financial risk posed by uncertainty of future energy prices which meant if prices plummeted, investors lost money. In guaranteeing prices, this scheme was highly effective, promoting the roll-out of solar PV and providing a financial underpinning to numerous community schemes (Nolden, 2013; Morris and Jungjohann, 2016).

The role of FiTs highlights how the design of energy markets and the rules of participating, be it in generating energy or using it, are central factors in the success of community energy schemes (Kooij et al, 2018). It also illuminates how the availability of these mechanisms is rooted in broader political aims. The support of community energy by David Cameron's coalition government formed an important part of its 'Big Society' approach to policy: in which communities were presented as being better-placed to serve their own needs and fill the gaps left by the state and financial markets. As the government's 2014 Community Energy Strategy put it: communities 'can often tackle challenges more effectively than government alone, developing solutions to meet local needs, and involving local people' (DECC UK, 2014: 7). Community energy projects were seen as one way to increase consumer choice to increase competition in energy markets. In 2013, the then Minister for Energy and Climate Change in the UK, Greg Barker MP (2013a), wrote: 'I want companies, communities, public sector and third sector organisations to grab the opportunity to generate their own energy and start to export their excess on a competitive, commercial basis. Not just a few eco-exemplars – the big six need to become the big 60,000' (Barker, 2013a). Barker's words were mirrored in a speech he gave on ground-source heat pumps in December 2013. In this second speech, Barker (2013b) argued that the expansion of the national energy sector to include 'the big 60,000' would allow government policy to unite ambitions to both get a better deal for electricity users and create a greener, more-local energy model. This chimes with the driving force behind many community energy projects today: ensuring benefits for both the climate and local residents.

Barker's narrative of consumer choice and 'opening up' the energy market later helped justify the removal of FiTs. In the UK, the scheme was reformed in both 2015 and 2016, after the election of a Conservative Party majority government in 2015. The scheme was closed to new applications in 2019. While subsidies and payments agreed before this date are honoured, the scheme has been replaced. There were several overlapping reasons for this move. First, FiTs were used by homeowners to install rooftop solar panels on their homes yet the declining cost of renewables (particularly of solar panels) meant that funding might be used more productively when it was channelled to larger projects. For the developers of these larger projects, declining costs meant that financial risk became more manageable without government subsidies. Second, price guarantees, locked in over 20 years, are expensive for a government to take on as a long term costs. The declining cost of renewables also meant that governments were underwriting long-term costs on technologies that were growing cheaper and cheaper. With the recovery of costs paid having to be found from somewhere, the costs of FiT's were passed on to consumers (Helm, 2017). The channelling of FiT funds to homeowners to be paid through other people's energy bills also has important energy justice implications, which I will discuss further in the next chapter.

In recent years, policies supporting this form of energy in the UK have shifted from supporting people-led projects to a greater focus on private companies and partnerships leading local energy networks (Devine-Wright, 2019). As a result, community energy projects can involve actors from both inside and beyond the local community and area, including funders, intermediaries, and private organisations. This has occurred alongside a move away from feed-in tariffs and towards competitive auctions, in which energy producers bid on available capacity and tendered contacts. The bid with the cheapest proposed cost usually wins. In the UK, a system called 'Contracts of Difference' (CfD) has been introduced. This scheme aims to protect project developers from future price volatility by paying a flat rate for the electricity a scheme generates over 15 years. Project developers submit sealed bids during an auction phase, based on revenues from selling electricity to the national grid. If successful, they enter into a contractual agreement with the Low-Carbon Contracts Company, a private company owned by the government, that agrees and assures the rates to be paid.

The phasing out of FiTs highlighted the dependence of many community energy projects on government policies that sheltered them from uncertainty and financial risks (Hannon et al, 2022). It placed many groups in a difficult position and slowed their expansion, as they sought to secure advances already made. Smaller community-centred energy projects must now compete in energy markets without protection from financial risk. This can limit their potential. In Germany, where FiTs were phased out in 2014, the pilots of a new auction system for solar PV between 2014 and 2016 were dominated by funding going to bigger energy companies. No energy cooperative was successful in its bid (Klessman and Tiedemann, 2017). An estimated 128,000 small-scale solar PV facilities will likely lose support between 2020 and 2025 in Germany (Sutton, 2021).

While state support might, in theory, be available to all, it is often out of reach of many due to the difficulties in accessing and securing it (Bomberg and McEwen, 2012). In the UK, the closure of the FiT scheme limited the potential of community-centred energy projects. The new system of CfDs does not currently apply to any of the areas that community energy projects might be financially able to work in. They remain focused on offshore wind, something that all communities would struggle to invest in, due to the huge costs required (Ambrose, 2020a). Onshore wind and solar projects compete for a separate pot of funds – the result of a government u-turn on blocking CfD subsidies for onshore wind (Ambrose, 2020b). The effort required to secure subsidies and financial support has stifled local-led energy projects (Boon and Dierperink, 2014). It will also likely discourage communities taking action into the future. This leads to a locked-in system, in which big energy organisations influence and limit the forms an energy transition might take and smaller, community-centred schemes can struggle to remain involved (Brisbois, 2019).

An example of this model can be found in Shetland, where the Viking Energy project was originally part-owned by the community through the local council. This stake was later passed in 2007 to the Shetland Charitable Trust, a form of sovereign wealth fund set up to use the proceeds of the oil industry for the community of the Shetland Isles. However, the scale and cost of the project (involving 103 turbines and a connection to the mainland) caused the Trust to withdraw. Having already invested £10 million, the Trust divested its 45 per cent stake in the project in 2019 (Platform, 2022). This decision was partly motivated by some local opposition (who appealed the decision in court in 2012) and the changing landscape of government subsidies and support, with the move to Contracts for Difference changing the financial risk that the Trust was able to take on.[9] Today, the project (under construction at the time of writing) is solely owned by Scottish & Southern Energy (SSE). The community retain representation on the management board and will receive £2.2 million a year in a community benefit fund (Shetland News, 2020). However, financial risk and uncertainty restricted the potential direct community involvement

The national policy context in the UK influences the possible forms that community-centred energy generation projects can take. Pitt and Nolden (2020) have defined two different models of community solar. In the 'local' model of community solar, the electricity generated could be transmitted to local users through a microgrid. This can provide extensive benefits to residents. Residents can pay a cheaper rate for their electricity, allowing savings on bills of up to 29 per cent (Pitt and Nolden, 2020). Cheaper energy can also be targeted to support those residents living in energy poverty. In Bethesda in North Wales (an area of high energy poverty), 100 households formed an Energy Local club in 2016 and partnered with Cooperative Energy (an electricity supplier) to buy electricity from a local hydropower plant (Possible, nd.). In the UK, the transmission infrastructure required for such a 'local' model of energy generation can often be prohibitive. This has led to a Private Member's Bill being introduced in the Houses of Parliament. The Local Electricity bill, first introduced in 2020 and, again with an amendment in 2021, calls for community energy generators to be able to provide electricity directly to local homes and businesses. However, without supportive government policy, doing so remains difficult for all community-centred energy schemes – due to the high upfront costs of building the transmission connections needed.

Many community-centred energy projects such as the wind turbine in Lawrence Weston adopt a different model. In the 'export' model, the electricity is sold onto the national grid directly, with the revenues invested in the community. The energy generation is owned by the community, allowing them to secure far greater benefits than they would from a community benefit fund provided by other projects. Yet, community energy programmes require significant buy-in, in terms of money, time (often voluntary), and expertise. All require supportive policies. Recent work has uncovered numerous barriers to community energy projects, including the poor coverage of financial support, limited access to land or buildings that most host community-owned infrastructure, the limitations of skills and time availability within community groups, difficulties in forming local partnerships with other organisations, and poor access to both energy markets and ethical finance (Hannon et al, 2022). The lack of supportive finance available means that the energy model remains dominated by institutional investors who prioritise large and low-risk projects. In the UK, the transformative potential of community energy has been held back by the way that the national energy model works: without concerted policy support, local schemes are unable to compete with the bigger energy companies. The result is that they become embedded in a system where their success cannot be scaled up or expanded. Big energy companies can fail, cut their losses, and focus elsewhere. Community-centred energy groups can't do that. The removal of FiTs has shown the need for a a safety net: without it, communities will be pushed out of an energy transition.

Local control

The scale of the energy transition requires supportive policy, long-term planning, and sustained financial support. A community-centred energy transition can be driven by other organisations and institutions that take a role and stake in new infrastructure and build and operate it for the benefit of the community. In Germany, policies to decarbonise the energy grid highlighted the importance of an active national government creating a context in which energy projects led by local authorities and governments might thrive. Formed of targets related to energy savings, emissions reductions and the expansion of renewables, the concept of Energiewende targeted wholesale change in the energy sector (Kemfert et al, 2015). While made at the national level, these policies took form in local and regional contexts (Moss et al, 2015). The era of Energiewende in Germany has coincided with an expansion of local and regional movements calling a greater control of energy grids.

This investment provided fertile ground for residents, communities, and local and regional governments to invest in renewable energy, forming new community-centred energy organisations. These processes generally represented a process of 're-municipalisation', in which public power at the local level is re-established over private energy utilities. In Germany, municipal, community and cooperative energy take numerous forms. Some are staffed by a handful of people, while others are more expansive. Stadtwerke are city-level energy utilities that provide electricity or heat to the population. In Munich, Stadtwerke München (SWM) is one of the largest municipal energy companies in Germany and provides enough energy to serve 95 per cent of households in the city (Weghmann and Hall, 2021). The Stadtwerke München employs thousands of people and holds an important financial capacity, having invested in offshore wind farms as well as in energy projects in France and Sweden (Morris and Jungjohann, 2016). These municipal companies raise funds via the sale of electricity and/ or services to local residents. Others also run telecommunications systems, public transport systems and sponsors local sports teams. The Stadtwerke in Vienna, Austria (following a similar model to those in Germany) manages 46 cemeteries in the city (Wien Stadwerke, 2020). These established companies are investing extensively in renewables; for example, Stadtwerke München has previously invested tens of millions of Euros into twelve solar PV parks (Effern, 2019).

These Stadtwerke exist in a particular context. Germany has a strong history of locally owned utilities and, while successive German governments have been reluctant to nationalise the energy sector, Energiewende policies created an important context in which the local groups and authorities were able to change the energy sector (Becker, 2017; Paul and Cumbers, 2021). The relative success of these policies in promoting localised responses is also linked to institutional factors, such as the provision of financial support in the form of long-term loans, with fixed (often low) interest rates that provided much-needed financial security (Nolden, 2013).

Processes of re-municipalisation are expensive. In Thuringia, a state in central Germany, 850 municipalities took majority ownership of its energy infrastructure and services in 2012, at a cost of €950 million (Paul and Cumbers, 2021). The takeover was funded by public loans – to be repaid by the municipalities, with the expectation that the venture will be debt-free by 2030 (Paul and Cumbers, 2021). This is one of the most expensive public takeovers in Germany and created important tensions between different municipalities and communities in Thuringia, due to different capabilities of being able to pay for it.

In Germany, public energy utilities have been important electricity providers for many and several communities and regional governments have seen an opportunity to regain control of energy and secure public ownership of energy suppliers (Becker, 2017). Examples of this can be found in Berlin and Hamburg, where communities sought to regain public and municipal control of energy networks. This was unsuccessful in Berlin, where a 2013 referendum vote failed to achieve the turnout of 25 per cent required as a mandate of action (but 83 per cent of those voting did vote for re-municipalisation). In Hamburg, the referendum resulted in a narrow majority of 50.9 per cent in favour of municipal ownership.[10] In 2014, the City of Hamburg reached a deal with Vattenfall to purchase the area's electricity distribution grid for €550 million. The energy grid entered public ownership in 2016.

Policies of Energiewende and re-municipalisation demonstrate the important role that national governments and institutions can play in accommodating local-led or community-centred energy schemes through supportive policy or finance (Kalkbrenner and Roosen, 2016). This process is important because it represents a reversal of previous policies of privatisation, returning goods and services to the public. Re-municipalisation doesn't just represent a regaining of control over energy – it is a reframing of what electricity is and why it matters, and the role that local and regional governments can take (Becker et al, 2017). In Barcelona in Spain, efforts to re-municipalise energy have been accompanied by aspirations for energy independence and addressing energy poverty (Angel, 2021). Re-municipalisation can also come to represent a turn inwards, with local government and institutions recognising that decarbonisation and climate solutions can come from within their communities (Haf and Robison, 2020).

The political and policy context underpinning the municipalisation of local energy systems is important. Pledges of nationalisation by political parties in the UK have been dismissed, criticised, and trivialised – symbolising the deep roots left by the privatisation of public utilities by the government of Conservative Prime Minister Margaret Thatcher (1979–1990). However, energy policy in the UK is not totally detached from the idea of government intervention in the energy sector. Bulb Energy was taken into public ownership after its failure in November 2021 and the 2022 Energy Security Strategy announced a new public body, the Future System Operator, responsible for planning and managing energy distribution to help fulfil climate action targets. The formation of this organisation involved taking control of some of the privatised National Grid's responsibility – and compensating it appropriately. Both moves were announced under the Conservative government of Prime Minister Boris Johnson. Both also suggest that government intervention in the energy sector at a time of crisis is possible when deemed necessary by those in power.

The key challenge facing many local authorities in the UK who might seek to follow the example of Stadtwerke is how the UK economy is highly centralised – both in terms of the geography of wealth (centred on London and the South East of England) but also where decisions are made. The devolution of policy-making powers to regional authorities has been slow and individual government departments continue to make wide-reaching decisions within a context of national priorities that don't necessarily translate to local or regional contexts. Local governments often have to bid to raise funds from central government pots of money. Important place-based factors and context can be lost, as regional leaders have to explain the importance of their plans to decision makers who may never have stepped foot in their town or city. This was labelled as a 'begging bowl culture' of funding by the Conservative mayor for the West Midlands, Andy Street, in early 2023 (Haynes, 2023).

In March 2023, the Chancellor of the Exchequer, Jeremy Hunt announced measures to hand control of regional finances linked to housing, skills, and transport to the West Midlands Combined Authority (led by Street) and the Greater Manchester Combined Authority (led by Andy Burnham) respectively. This represents an important shift in how public funds are managed, giving greater control to regional leaders to invest in community-centred and place-based initiatives. However, it must reach wider. While centrally-administered funding pots are available, local authorities in the UK have been highly impacted by government policies of austerity. Funding for local government between 2010 and 2018 was reduced by 49.1 per cent in real terms (UK Parliament, 2019). This has had important consequences for the capacity of local authorities to provide basic services or pursue broader policy change. There is only so much that local authorities can do at the current levels of funding and capacity. With the planning and development of renewable energy infrastructure existing as a discretionary part of any local authority's responsibility, it often falls down the priority list in the face of squeezed budgets (Sugar and Webb, 2022). The authority given to regional leaders in the West Midlands and Greater Manchester must be extended far wider.

Rather than buying energy networks out of private hands, some local councils in the UK have set up companies to compete in the energy market. One important example of such a move is Bristol Energy – which was founded in 2015, at a time when numerous local authorities in the UK were seeking to enter energy markets. Owned by Bristol City Council, the company took both commercial and residential customers and was advertised as an energy supplier that was good for both communities and the climate, with the city's population identified as its core customer base. By 2019, it had 165,000 residential customers (Bristol Energy, 2019). However, the company soon encountered difficulties – particularly after Spring 2018, when it lost the contract to supply Bristol City Council (its owner) to British Gas, due to strict rules on procurement (Wilson, 2018). While this contract was later re-secured, the company entered a period of uncertainty, making financial losses and seeing two chief executives leave in quick succession. Cash injections from public funds were required but, eventually, the financial losses of the entity became too much for Bristol City Council to bear. In 2020, Bristol Energy was broken up and sold.

A report on Bristol Energy, written for Bristol City Council by the auditors Grant Thornton, found several failings of governance, including a shallow appraisal of options at the time the company was set up and missed opportunities for the Council to cut its losses (leading to an inflation of costs to the taxpayer from £19.8 million to as much as £43.8 million) (Postans, 2021). At the time it was set up, the intention had been to introduce both an energy services company and an energy supplier. While the former would have been a business that provided support for energy savings and retrofitting, the latter provides energy to consumers only. Only the latter was founded in Bristol Energy, and there is little information available on why this was the case. Such a decision represented a divergence from policies of local and city governments elsewhere, with other local authorities teaming up with private suppliers (such as London Power, a partnership between the Mayor of London and Octopus Energy).

The municipal ownership of energy networks can be transformative, providing new sources of local revenues and allowing for more decision-making capacity to accelerate local climate action (Sugar, 2021). Bristol Energy piloted an innovative business model, offering a flat rate tariff based on the temperature of the home, rather than billing per unit of electricity used (Brown et al, 2020a). However, the story of Bristol Energy has primarily become an example of how difficult it can be to 'compete' in the current energy model. Bristol Energy failed because it was set up in the belief that, as a municipally owned energy provider, it could directly compete with bigger companies in a volatile market. This was a misperception in two parts. First, in the view that it would be able to entice customers away from competitors due to either their Bristolian links and identity or through cheaper bills. Second, the belief that not having to pay annual dividends to shareholders would allow for cost savings to be passed onto customers and attract more customers. However, due to their size, this model was less able to enter zero-sum price wars with bigger, investor-owned utilities that absorbed financial losses from undercutting smaller challengers to gain and retain customers and ensure their market position.[11]

Bristol Energy was an ambitious policy and an important lesson lies in the prioritisation of forming an energy supply company over providing energy services and advice only (Grant Thornton, 2021). This doesn't necessarily need to be the case. Instead, local governments can act as enablers, providing direct support to communities seeking to generate energy, and as fulcrums for the building of a broader tapestry of local energy. Bristol Energy did go some way to doing this, providing favourable conditions and payments to small energy producers in the local area (Brown et al, 2020a). As a result, its story still signals a potential of fusing together local economic aspirations with a new energy model.

Rule 3: Foreground community-centred energy schemes in local economies and wealth building

Community and public-owned electric projects can become key lynchpins in energy transitions, both linking decarbonisation to local communities and providing benefits to them. In doing so, they can go some way to addressing and avoiding distributive and procedural injustices. They can also form close relationships with local organisations, putting revenues to use to help others in the community, be it in terms of supporting education or encouraging broader changes to reduce greenhouse gas emissions. The organisations are out there: ranging from the Brighton & Hove Energy Services Cooperative to the Abernethy Trust Hydro and Electricity Scheme. However, these community energy ventures can be (and have been) held back by a national energy model that continues to serve bigger companies.

While decentralised, community energy may not be able to provide the scale of energy transition needed, it highlights the importance of national energy policies to make space for community-centred transitions. Some have highlighted the possibility of these schemes being partnership with local authorities, public bodies, and institutions (Hannon et al, 2022). This speaks to a 'new municipalism', or the democratic transformation of the local economy through a more-active local government (Thompson, 2020). However, municipal approaches to energy can also lead to injustices that all involved should be mindful of. In some places, municipal energy projects has been used as a vehicle to further individual political careers and generate local revenues that are shared unevenly, often primarily benefiting town or city centres rather than those living elsewhere (Rudolph et al, 2018;

Emelianoff and Wernert, 2019). Local authorities exercising greater control over energy projects in an area might also crowd out smaller, community-led projects, further excluding some groups from energy transitions (Knox et al, 2022). These potential impacts necessitate linking municipal energy projects to broader policy goals.

In the UK, there are important cases of local authorities seeking to create the contexts in which other institutions and communities can exercise influence and control over local economies. UK policy has shifted from community energy towards a focus on 'local energy', in which local groups and organisations join forces with private interests (Lacey-Barnacle, 2020). In North Ayrshire, the local council have built council-owned renewable energy facilities but rather than selling the electricity on the market, they have used it to power council-owned buildings. A planned council-owned solar farm in the area is anticipated to fulfil 34 per cent of the council's energy demand and generates a financial surplus of £13 million (Cullinane, 2021). In 2022, new projects had been planned, including three wind turbines and another solar farm (Inglis, 2022). Income and cost savings from these will be used to fund programmes to address energy poverty in the area.

Energy transitions have an important role to play in local economies and supply chains, be it involving private or public-owned institutions (Leffel, 2022; Sapir et al, 2022). There are lessons to be learnt from the 'community wealth building' approach of local economies in the UK. Such as in Preston, where the City Council has adopted an economic model that seeks to leverage its role as a major buyer of goods and services to stimulate local economic change, create new jobs and generate new revenues for local businesses. This 'Preston Model' has worked with other anchor institutions – such as the National Health Service and schools, universities, and colleges – to increase the buying of goods and services from the local economy and encourage suppliers and local businesses to invest in training and job-creation. In the wake of the 2008/9 economic crisis and resultant recession, this strategy proved transformative for the local economy. Between 2012 and 2017, the spending of these organisations with Preston-based suppliers increased from 5 to 13.2 per cent – increasing revenue for local businesses by close to £75 million (Cannon and Thorpe, 2020). This came to support 1,648 jobs in Preston and a further 4,500 jobs across the Lancashire region (Cannon and Thorpe, 2020).

Jamie Driscoll (2021), the Mayor of the North of Tyne Combined Authority in the north-east of England, has argued that regional wealth-generation can help reduce inequalities and skills gaps which would, in turn, lead to further economic benefits. A key route for this could be found in Community Municipal Investments (CMIs) – bonds that might be issued by local and regional authorities and bought via crowdfunding platforms. Local residents and companies can buy these bonds, with the money raised than invested in projects and schemes that deliver local policies and priorities: such as for a just energy transition (Davis, 2021). Investors don't necessarily take on the financial risk associated with the funded projects which, instead, become linked to the council's ability to raise tax long-term revenues. Within such schemes, local authorities can become a key driver for decarbonisation in the UK: with it estimated that, if all 404 councils offered CMI schemes, over £3 billion might be raised (Davis, 2021). In Warrington, Cheshire, a CMI scheme in 2020 attracted 500 investors, investing an average of £1,921. The funds were used to fund two large solar farms and battery storage facilities (Davis, 2021). A council-led pilot in West Berkshire the same year saw £1 million raised. Such schemes can also lead to a greater awareness of decarbonisation amongst residents, and an increased transparency of council policy: in both cases, investors chose to re-invest interest payments back into the scheme.

Municipal governments and local companies also need to show sustained and comprehensive commitment to community-centred renewable energy schemes. It is a sad irony that the collapse of Bristol Energy was, in part, the result Bristol City Council, moving its own energy contract elsewhere due to procurement rules. Within community wealth building approaches, the land, buildings, and capabilities held by local governments are not seen as assets that might be bought or sold – instead, they are used for the common good (Cullinane, 2021). Community-centred energy should be encouraged to play a role in community wealth building and become linked to a broader network of public banking, local innovation, and community procurement programmes (Hanna et al, 2022). As in the 'Preston Model', large local employers should be encouraged to use their economic position to both fund renewable energy schemes and to adapt their procurement rules and processes to encourage and support community energy programmes. There is a need for an approach that treats renewable energy as a 'public good' in which community-centred initiatives have a role in an energy transition and local wealth building and that local authorities and companies work and invest nurture such a capacity. This can take numerous forms, including improving homes through energy efficiency retrofitting and job-creation: topics that I discuss in the next two chapters.

Empowering a community through energy policy requires the understanding and elevation of a community's own goals and voices (Coy et al, 2021). Community energy, such as the wind turbine at Lawrence Weston, represents a form of public wealth, in which electricity is generated and sold and the revenues are put to use in supporting local residents. In Lawrence Weston, the proceeds of the wind turbine will be returned to the community. There are hopes that this will be used to open a skills and training facility, in which residents are given the opportunity to develop the new green skills that decarbonisation requires. This will be transformative in a community where over one third of residents don't have formal education qualifications and which has, in the past, lost its local college (ALW, 2022). Building renewable energy infrastructure and services provides a route to addressing energy poverty by helping keep the lights on and their homes warm. This requires an understanding that energy transitions can be restorative, as well as transformative. With the most vulnerable facing increased energy insecurity, it is necessary for a just energy transition to seek to make their lives better. Such an approach of restorative justice begins in people's homes.

6

Home

As I walk past tenement buildings and takeaways in the East End of Glasgow, a cyclist with speakers kicking out drum and bass tears past. We both pass stores with their shutters down and communal gardens with the morning's laundry hanging in the sun. This part of the city experienced much change in the latter half of the 20th century. The Glasgow Eastern Area Renewal process, launched in 1976 and ending in 1987, saw the neighbourhood's derelict houses and industrial buildings demolished, tenements renovated, and coal fireplaces replaced with gas central heating. Such change continues. In front of me, at the point where Shettleston and Wellshot Roads meet, stands Cunningham House: a five-storey building fused to the Carntyne Old Parish Church. The church, built in 1893, has stood as a landmark in this community for generations. Today, it holds its original form, but past and present merge into one. Restored lancet windows hold double-glazing. Stained glass tints light flowing onto new kitchen fittings. Inside are 19 new homes for older people. All dwellings within the two buildings are designed to meet the Passivhaus standard, which aims to limit how much heat leaks out of a home and ensure a consistent and comfortable temperature inside. Rain or shine, warm or cold. This both reduces the building's emissions and the energy bills for the people who live in these homes. Cunningham House is the first Passivhaus development in Glasgow, but there are others. Another scheme has been built in Nitshill on the southside of the city, with 178 new energy-efficient homes on a brownfield site available for social housing, rental, or sale.

In an era of spiralling bills, energy efficiency measures can be transformative for many – such as the vulnerable older people at Cunningham House. Energy poverty is one of the biggest challenges for many in the community today. In the UK, energy poverty is understood as when household energy bills cost so much that, after paying them, a household's leftover income is below the official poverty line (set at 60 per cent of the national median income, which is £31,400) (BEIS, 2022c; ONS, 2022a). In 2022, an estimated 3.16 million households (or 13.2 per cent of the total) were understood have been living in energy poverty in the UK (BEIS, 2022c). Of the households in Glasgow, 34 per cent are estimated to have lived in a period of energy poverty at some point between 2012 and 2014 (Glasgow City Council, 2016).

It was estimated that energy price increases would plunge an estimated 8.2 million homes into energy poverty over the winter of 2022/23 – a third of all households in the country (Morison, 2022). Despite government financial support, the 2022 energy crisis pushed many who were unable to pay their bills onto pay-as-you-go energy meters, where you pre-pay for your energy via topping up a card or account. It was estimated that energy companies moved 600,000 people onto pre-payment meters in 2022 alone (Citizens Advice, 2023). For many, these meters represented a 'disconnection by the back door', with an estimated total of 3.2 million people running out of credit on their meter and going without energy at some point in 2022 (Citizens Advice, 2023). While government policy must act to reduce such vulnerability in the short term it should also focus attention on how this energy poverty might cluster in neighbourhoods like Shettleston and other low-income or vulnerable communities across the country.

Our homes are where we spend our lives. They are where we eat, rest, and close the door away from the world. They are also where we are most aware of our energy use. Everyday energy demand is intricately connected to how we live our lives and who we live them with. Yet, household energy demand is often overlooked in the policies decarbonising our energy grid. UK policy primarily focuses on building more energy projects to boost the supply of electricity, rather than thinking about how we might change patterns of demand. To keep global heating below 1.5°C, we also need to reduce how much electricity we use (Diesendorf, 2022). This chapter argues that a just energy transition needs to be brought closer to home. In a time of energy crisis, one of the quickest and most effective things to do to reduce people's energy bills is rolling out home insulation and energy efficiency measures and, in doing so, limiting how much energy our homes need to use. These measures will reduce energy poverty and increase energy security, at the household, community, and national levels. Energy transitions make our homes warmer and bills cheaper. They can allow for policies to focus attention on where it is needed most: cold homes, vulnerable households, and helping those excluded from previous schemes.

Keeping the lights on

The inclusion of policies that make homes low-carbon, warmer, and cheaper to run holds a significant role in a just energy transition. A lack of access to energy represents injustice in terms of distribution, procedure, and recognition. A household living in energy poverty (and its impacts) might see the routes to address this problem closed off and the causes of the issues they face misunderstood or poorly represented by those making decisions (Walker and Day, 2012). Addressing issues of energy poverty can also promise a form of restorative justice, in which decarbonisation can improve people's lives, particularly those who are already vulnerable or excluded, marginalised, or ill-treated by previous policies. The idea of restorative justice originally emerged from the legal sphere, in a process that aims to both punish an offender and repair the harm done to a victim. For environmental and climate policy to be just and restorative, it must seek to both reduce impacts and give those impacted a better life. The emergence of just transition narratives from the trade union movement holds an important element of restorative justice, with it based on ensuring that workers displaced from fossil fuel industries are supported and given new opportunities. Restorative justice approaches in an energy transition call for policies that address past damages to individuals, communities, and the climate, and sees decarbonisation as an opportunity to provide and support better lives for those impacted and marginalised by previous policies (McCauley and Heffron, 2018; Hazrati and Heffron, 2021). In the United States, Shalanda Baker (2019; 2021) has argued that policy makers and researchers should always assume that energy systems have harmed low-income and Black communities: be it through direct pollution, creating a dependence on dirty jobs that damage health, or through the development of energy sacrifice zones. To mitigate this, policies shouldn't just aim to address past injustices but also to disrupt the practices and activities that lead to them, such as through new ownership structures and approaches (Baker, 2019).

A restorative justice approach to household energy transitions in the UK requires action on energy poverty. A just energy transition must address the *energy trilemma*: to ensure energy availability (linked to the security of supply), accessibility (ensuring that everybody can financially afford energy), and sustainability (reducing emissions) (McCauley, 2018). Experiences of this trilemma differ across households, communities, and regions due to the different factors that might influence energy poverty. An energy crisis impacts the most vulnerable. In the winter of 2022/23, spiralling energy bills disproportionately impacted those living in poorly insulated homes, which leaked heat quickly and required more energy to keep them warm.

Our homes are also where inequalities and vulnerabilities take individual form. Energy poverty is no different. Energy poverty often intersects with geographical vulnerability, as well as structural inequalities, access to technologies, and our lived experience (Bardazzi et al, 2021; Willand et al, 2021). It is more common in urban areas, where those affected are often impacted by other inequalities such as a lack of access to transport (Robinson, 2019; Robinson and Mattioli, 2020). In the UK, single-parent households and Black and Ethnic Minority families are some of those more likely to face energy poverty (BEIS, 2022c). Energy poverty is more likely to impact people with disabilities, exacerbating pre-existing illnesses and symptoms and complicating the use of at-home medical equipment, such as that used by at-home oxygen therapy (Snell et al, 2015). These technologies rely upon uninterrupted supply, making many households energy dependent for their health and survival. Courses of hospital treatment also increase household demand, with those undergoing chemotherapy or radiotherapy due to cancer diagnosis requiring a warmer home to ensure their comfort and aid recovery. People with disabilities are also vulnerable, due to having to use electricity to charge mobility aids, such as wheelchairs, spending more time at home, or seeing income from disability benefits disappear in the face of rising prices. The significant links between health and warmth can be seen in actions by the UK National Health Service, implemented in late 2022 to prescribe heating to patients who couldn't afford to pay their energy bills. Prescriptions particularly targeted those on lower incomes with respiratory conditions and were made as a form of pre-emptive action to reduce future health problems (Turner, 2022).

The injustice of energy access means that any policies seeking to address energy poverty need to recognise and address inequality, rather than treat the problem in blanket terms. However, the current energy model restricts the possibilities of such an approach. Patterns of ownership in the energy sector have led to energy efficiency policies and schemes being limited by the ways that companies or policy makers understand the problem or the technologies available (Bouzarovski, 2022). Many policies and schemes provide some support but have failed to recognise how energy consumption and energy poverty are both place- and context-specific. A desire for profit within the shareholder-driven electricity sector can also lead to energy companies overlooking energy efficiency measures, due to their relatively limited potential for future financial returns (Newell, 2021). While energy companies highlight policies and processes that might support those continuing to pay their bills, they often overlook the element of restorative justice that can make measures and interventions transformative for many.

Energy poverty is far from just a number on a government spreadsheet. It is linked to a person's integrity, and their ability to live a good life and feel fulfilled. A cold house, a meter that is always running low, or a disconnected gas supply – all have consequences for people's lives. Over 40 per cent of fuel poor households in the UK in 2020 had children living in them (BEIS, 2022c). That's 1.3 million homes of young children that may be cold or damp, leading to health impacts. In November 2022, the tragic death of two-year-old Awaab Ishak in Rochdale, Greater Manchester, was attributed a respiratory condition caused by extensive mould in his home – a problem caused by energy inefficiency and damp. The cold can also kill: it has been estimated that 8,500 deaths in the UK a year might be attributed to living in a cold home (National Energy Action, 2020). Very cold winters can lead to tens of thousands of deaths, particularly of those over the age of 65 (Age UK, 2016). In total, an estimated 2.5 million people may have died in the UK since 1950 due to cold homes (Age UK, 2016). Energy poverty is not an abstract problem – it is a lived and everyday threat to people's health and lives.

All forms of energy poverty are dynamic. Things can improve or get worse for households across the calendar year. Some months are harder than others. Households may move out of energy poverty seasonally (due to changing needs) or suddenly (after receiving a particularly high bill or losing work) (Robinson et al, 2018). Others may find themselves in a state of energy poverty for the longer-term, due to certain demands of use, sustained unemployment, or other factors. In many cases, energy poverty can come to represent a vicious cycle in which vulnerability and the opportunities to address it are dictated by variables outside of their control (Bouzarovski and Simcock, 2017).

Rising energy bills will hit those on the lowest incomes the hardest. This is due to the disposable income available to absorb price rises. Poorer households often spend up to three times as much of their earnings on energy bills as the richest households in the UK (Marshall, 2021). As energy prices rise, those on lower incomes face spiralling costs that richer households are better able to weather. Rising energy costs in 2022 equalled an average of 6 per cent of the disposable income of the poorest 10 per cent in the UK, but only 0.75 per cent of the disposable income of the richest 10 per cent (NEF, 2022). This is despite the fact that these poorer households often use less electricity to begin with.

Issues of household energy insecurity are caused by the interaction of household income, energy-inefficient homes, and energy prices. Homes that leak warmth are important drivers of energy poverty and the UK's housing stock is the oldest in Europe: 20.6 per cent of it was built before 1919. The archetypal terraced homes that many live in were mostly built 150 years ago (Piddington et al, 2020). The age of these buildings mean that they often leak heat far quicker – through gaps in windows, doorframes, and walls. In the UK, our homes leak an average of 3°C of heat over 5 hours, compared to 1.5°C in Italy and 0.9°C in Norway (tado, 2020). This problem is well-spread across the country: median energy efficiency ratings and fuel costs are relatively consistent across the regions worst hit by energy poverty (such as the West Midlands) and those less-affected (such as the Southeast) (ONS, 2022c). It remains a national problem needing a national solution.

These issues can, at least in part, be mitigated by retrofitting – a term that is used to describe home improvements that change and reduce the energy demand of a household. This involves interventions to increase the energy efficiency of a building. There are numerous measures and technologies to choose from, such as roof or wall insulation. These can be relatively low hanging fruit, such as the replacement of windows with more-modern double or triple glazing. Measures can also be more noticeable, like rooftop solar panels.

A restorative justice approach to energy poverty illuminates the numerous benefits of energy supply and security. These advantages included those linked to heating, cooking, lighting, hot water, connectivity, warmth, the running of medical equipment, household security, improved air quality, reduced respiratory and heart illnesses, better mental health, and a reduced number of winter deaths (IEA, 2022b). These benefits (and the changes that allow them) will become even more important in household energy transitions, due to their sitting at the intersection of decarbonisation and the cost of living. However, despite previous government policies, these advances are primarily enjoyed by those who can afford them and remain out of reach of those most in need.[1]

A problem with prosumers

A focus on the household in energy transitions necessitates people being given the opportunity to take a more active role in shaping, changing, and reducing their energy demand. This can take various forms. From the small act of switching the plug off at night to more-drastic changes to your home, such as installing rooftop solar. As of 2021, over 61,000 properties in the UK had solar panels installed, taking the total number of solar-roof-topped homes to over one million (Reid, 2022). Taken together, these panels can provide around 3 GW of electricity, roughly the same as the planned Sizewell C nuclear plant in Suffolk, England. This installation of rooftop solar is often described using the label of 'prosumerism': a fusing of the words 'consumer' and 'producer'. A 'prosumer' simultaneously generates electricity via their solar panels, uses it in their home, and sells the surplus to the electricity grid. While group-buying schemes exist, the installation of rooftop solar is mostly arranged on an individual basis: in which a homeowner makes their own personal calculation of investment, returns, and desire to select whether to install solar panels on their roof. When circumstances change, the calculation might be altered also. This was evident in the boom in demand for solar panels in 2022, with increasing energy bills leading to more homeowners looking to make such an investment. That summer, enquiries about solar panel installation increased tenfold, with an estimated 3,000 homes having the technology installed every week (Read, 2022).

People taking an active role in changing their own energy footprint has often been described using a language of 'energy citizenship', in which individuals become responsible for their role in decarbonisation. A language of 'citizenship' in an energy transition presents people and households as responsible for their energy demand, and the emissions associated with it.

There are number of problems with this. First, it can come to represent a process in which government policy relies upon individual and local work and efforts in decarbonisation agendas, without necessarily creating the policy and technical terrain for such projects to easily find success. The transformative element of a just energy transition becomes the responsibility of the individual, not the government or energy companies who can, instead, focus on more-limited measures. A focus on 'prosumerism' and 'energy citizenship' also differentiates those who participate in an energy transition from those who do not or cannot. In doing so, it distinguishes an apparent 'good' energy citizen from a bad one, with the opportunities for such 'citizenship' skewed towards those who can make such a change or investment.

A language of 'citizenship' has always represented some form of exclusion and the use of this terminology in the current energy model is no different. In a political climate that has building walls and turning migrants back at the border being prominent narratives, the terminology of 'citizenship' can function to exclude. In an energy transition, the term ignores the political and economic factors that restrict people from being able to participate (Lennon et al, 2020). People being time-poor, lacking the ability or confidence to phone energy companies to negotiate tariffs, and language barriers might restrict engagement yet are ignored in an energy model that celebrated 'good' energy citizenship (Reames, 2020). Those who can install solar panels and retrofit their homes are presented as success stories and as good citizens doing their bit. Those who aren't are excluded, with little done to recognise or address the reasons for such exclusion.

The treatment of residents and households as exclusively energy consumers risks overlooking that not everybody has the resources needed to control their energy usage. It can also neglect the needs of specific groups, such as the elderly or the chronically ill. Many household energy policies still adopt the 'information-deficit' model, which assumes that people need to be informed, educated, or encouraged to make changes and, therefore, focus on raising awareness of energy demand. Interventions like this include: Energy Performance Certificates, which educate us on how much energy appliances might use or how leaky or efficient our homes might be; and smart meters that are designed to raise awareness about how much energy we are using in real time. Both attempt to nudge energy users towards certain interventions and forms of behaviour change.

However, household energy changes are not only driven by awareness and knowledge. The capacity of somebody to act and make changes to energy use, even with a knowledge of what might need to be done, is not shared equitably. There are many variables – such as available finances, time, levels of education or organisation – that lead to people approaching challenges in different ways. The changes required to make a home more energy efficient might be unaffordable. They might fall down the list of priorities as new costs and problems arise. Those in our communities who are on the lowest incomes often live in housing of the lowest quality and energy efficiency, or have limited housing security – frequently moving, lacking options, or facing higher housing costs (Boardman, 2009; Lewis et al, 2019). Addressing energy poverty and injustice requires the recognition that those vulnerable to it may be the least likely to have the resources needed to change things (Walker and Day, 2012).

In the UK, the people most likely to consider rooftop solar are a diverse group, in terms of both gender and ethnicity; but they are primarily younger (less than 35) and have a higher earning potential (BEIS, 2021a). Many have just bought (or will soon buy) their first home, signalling the importance of energy efficiency for new homeowners in the UK (BEIS, 2021a). However, measures like rooftop solar panels remain skewed towards those on higher incomes. In the UK, those with rooftop solar tend to live in homes of a higher value and have higher credit scores, with both rooftop solar and domestic wind technologies primarily adopted in high-income neighbourhoods (Sovacool et al, 2022; Stewart, 2022b). The links between rooftop solar and high incomes are not restricted to the UK. Research across four states (California, Massachusetts, New Jersey, and New York) found that nearly 90 per cent of rooftop solar installations in these regions have been on houses with annual incomes of over US$45,000 (£37,000) (Reames, 2020). Households with an income of less than $40,000 (£33,000) account for less 5 per cent of those with rooftop solar in the US, despite making up 40 per cent of the total number of households across the country (Reames, 2019). In California, half of the state's household solar is on top of the homes of the top 20 per cent of earners and only 4 per cent is found on the roofs of the poorest fifth (Hausfather and Stein, 2022).

Solar panels being on some roofs but not on others is, in part, linked to the policies set up to encourage people to install them. Accessing government schemes to support new energy measures takes time, energy, knowledge, and money (Sunter et al, 2019). Each rooftop project is unique, requiring individual design, construction, and payment. To offset these individualised costs, governments have used the Feed-in Tariffs (FiTs) (discussed in Chapter 5 and also called 'net metering' in the US) to support homeowners taking on the economic risks associated with rooftop solar investments. The way that FiTs and other government schemes are set up perpetuates energy injustice: households secure financial support to install rooftop solar on their homes, reducing the upfront costs that they pay, even though they may be better able to bear these costs than others.

Households with higher incomes are more likely to use these supporting schemes to install rooftop solar (Sommerfeld et al, 2017; Stewart, 2022b).

These households are then able to make money from the electricity their new panels generate in two ways: both through the savings on their energy bills (through the electricity created on their roof) and the revenues from selling surplus electricity (through prices guaranteed by the feed-in tariff scheme). The upfront investment is covered by the taxpayer and the payments made for the electricity generated on somebody's roof are paid for by the energy grid, leading to higher energy prices for other billpayers. In short, people can pay twice to financially support somebody else having rooftop solar – first through their taxes being used to support subsidies, second through increased electricity bills. A new class of 'renewable rentiers' emerges, who benefit from government subsidies both in paying less for the up-front investment and continuously gaining revenues from selling electricity to a national grid (Huber and Stafford, 2022). This creates an 'inequality trap' in which government subsidies support wealthier homeowners, widening the gap between these residents and those on lower incomes in terms of both energy security and household incomes (Stewart, 2022b).

Taking advantage of government schemes also requires the autonomy to be able to make changes to your home. Subsidies are often only accessible to those who own their homes and have higher incomes. This immediately limits the opportunities of many, particularly those renting their homes and those who are struggling to make ends meet. Energy efficiency retrofitting can be experienced differently by renters compared to homeowners. Landlords hold extensive power over the properties that they own and rent out, with tenants requiring their permission to install new measures. Strained relationships and absent landlords restrict the opportunities for tenants to make positive changes and enjoy the benefits of retrofitting (Hernández and Phillips, 2015). Landlords may also be less likely to invest in such measures, due to the financial benefits being primarily enjoyed by the tenant (Brown et al, 2020b). In turn, if a property owner does install these measures, tenants may see rent increases. Lastly, with landlords able to evict tenants with relatively limited notice, there is no real incentive for those renting (and living) in these homes to invest money or time in installing new energy measures.

In the UK, a near-necessity of homeownership in accessing rooftop solar excludes at least 4.5 million households from the technology and its benefits (ONS, 2018). This could be as many as 13 million people and one in four families (Generation Rent, nd.). Homeownership in the UK is heavily skewed by age and increasingly so. In 2017, half of those in their mid-30s to mid-40s had a mortgage, compared to two-thirds of the same group in the late 1990s (ONS, 2020). Close to 75 per cent of people over 65 own their home outright (ONS, 2020). Today's young adults are less likely to own their home at any age than those born five or ten years earlier (Cribb et al, 2018). With those most likely to consider rooftop solar in the UK being under 35 and just starting to own their homes (BEIS, 2021a), the problems facing this generation in home ownership will likely restrict the expansion of this technology. A lack of access to owning your own home also has broader consequences: the group facing the highest levels of energy poverty in the UK are those living in privately rented homes (BEIS, 2022c).

Policies that have supported the boom of rooftop solar have often only covered part of the costs of installation. Group-buying schemes, such as the Solar Together scheme, in which local councils organise residents to bulk-purchase solar panels, can reduce the costs of installation to around £4,000 per household (O'Shaughnessy, 2022). This scheme provides an important model for supporting the roll-out of rooftop and solar and has had some significant success: helping thousands of households access the technology. However, rooftop solar remains out of reach for many. Installing solar panels could cost anywhere between £2,000 and £8,000. This money needs to come from somewhere and many people and households are unable to take advantage of such schemes. Some may never dream of having this much money available. Other long-term expenses such as mortgage repayments or childcare or eldercare responsibilities limit the finances available for households to pay for solar (Sovacool et al, 2022). The upfront costs of the technology remain prohibitive for many.

People are also only able to get rooftop solar panels installed on their roofs if they own that part of their home. This complicates the process for those living in flats and apartments. In England and Wales, an estimated 5.4 million households live in flats, maisonettes, or apartments – 21.7 per cent of the population (ONS, 2023). These households are unable to exercise any control over their roof space and, as a result, are ruled out of any expansion of rooftop solar. This exclusion has an additional layer of injustice: with 4 per cent of those living in lower-income areas living in high-rise blocks, compared to less than 1 per cent of those living in the least economically deprived areas (MHCLG, 2019).

Without policies aimed at restorative justice, the benefits of technologies, like solar PV, will remain out of reach for many. This has resulted in solar PV being seen as a technology that is only used by the rich and which is out of reach of many households. Even when tenants in social housing are offered solar panels, they might not see the technology as something for 'them' – instead perceiving it as 'posh' and primarily for wealthier households (Fox, 2018). Communities in the United States have even lobbied against rooftop solar schemes, supporting big energy companies in efforts to policies that support the roll-out of solar PV. This is likely a consequence of how a lack of regulation of solar schemes in the United States leading to many low- and medium-income households falling victim to predatory finance mechanisms. This includes those signing up for rooftop solar projects sold by door-to-door salesman who charged extensive interest on loans for panels that didn't work (Burns, 2021). Some Black communities in Florida and Illinois supported efforts to restrict the spread of rooftop solar in their states, seeing this technology as out of their reach and characterised by an injustice of uneven access. In doing so, they, perhaps inadvertently, provided support to incumbent energy companies that see rooftop solar as a threat to their business model (Penn, 2020).[2]

Rooftop solar is changing the energy grid. However, this can in often be in an unjust way, unless there is a sustained focus on using these technologies to support people in energy poverty (Lacey-Barnacle, 2020; Stewart, 2022b). This poses an important problem for future policies. They must seek to support decarbonisation and address energy poverty and injustice by prioritising support for those who would otherwise be unable to benefit from such schemes.

From the ground up

A notion of community is important in retrofitting schemes, often due to the role of trust. Researchers have shown that, if somebody distrusts their energy company, they are more likely to consider installing rooftop solar and seek to ensure their own autonomy from the energy grid (Horne et al, 2021). When thinking about solar PV or energy efficiency measures, many people trust their neighbours and communities more than they would somebody else. As a result, the uptake of solar technologies in neighbourhoods is often linked to local networks. Referrals from neighbours or other people in the communities are one of the lowest-cost methods of expanding rooftop solar. Word of mouth is an important factor in people taking-up energy efficiency measures, with neighbours learning from one another and trusting their judgement and experience (Snell et al, 2018). As neighbours learn about the cost savings those living next door have made, they might be encouraged to follow a similar path. Referrals within communities have accounted for some 40–50 per cent of installations in California (Sigrin et al, 2022). This happens across the political and ideological spectrum, with residents sharing information about schemes and benefits across other divides (Mildenberger et al, 2019).

Community-centred initiatives have shown themselves to be well-placed to encourage and harness such trust in an energy transition. In Sydney, Australia, the Voices of Power project provides targeted energy training, in which community leaders in the city are trained in not only understanding the potential of solar PV and retrofitting but also in how they might encourage others in their community to adopt these new technologies. [3] These leaders speak to their community (including Vietnamese, Chinese, Indian, Jewish, and Filipino communities in Sydney), hear and collate their needs, and devise ways in which new energy measures might promise wider benefits. This scheme reaches out to communities who often feel detached from those who make policy decisions and gives them space and encouragement to devise their own strategies, identify funding streams, and explore possibilities. In doing so, the Voices for Power project opens the possibilities for an approach of restorative justice, in which different communities not only influence the forms that an energy transition might take but can link it to their own experience and histories. Other local energy schemes can be focused on supporting low-income households and communities, with retrofitting and rooftop solar programmes addressing issues of energy insecurity and poverty through collaboration with communities to understand where the greatest need is. For example, in California, the Greenlining Institute works directly with communities to align electrification schemes to residents' needs (Bouzarovski, 2022).

Community-centred solar approaches can also allow for projects to find and use new locations beyond rooftops to create a resilience of local energy systems in the face of rising energy insecurity. Many people who can't have rooftop solar can be included in this energy transition through community-owned solar arrays that provide electricity to residents. Solar arrays could also be placed on the roofs of public buildings or on any building with enough roof space. Key local institutions and spaces, like schools, often require additional funding and have looked to solar panels to generate new funds by selling the electricity generated. In England, community organisations have been set up to support schools to install rooftop solar. The Schools Energy Cooperative, launched in 2014, has raised shares from members to install arrays on the roofs of over 90 schools across the UK. The cooperative brings schools together with energy sector intermediaries, local groups, and technicians and pots of funding to install solar panels on a school's roof, free of charge. All profits raised are, in turn, returned to the schools themselves. Schools that have benefited from this scheme can be found across England, including in the towns of Middlesbrough, Bournemouth, and Rochester. At other schools, the cooperative model of this venture is analysed and studied by students in their business studies classes, allowing them to understand how the solar panels above their heads are funded and managed (Schools Energy Coop, nd.). Elsewhere, solar panels installed by the organisation power hydro-pools, sensory rooms, and therapy equipment used in schools that serve pupils with complex learning difficulties (such as autism spectrum conditions or sensory impairments).[4]

New energy infrastructure should not be only built on the rooftops of public buildings like schools. Unused land in neighbourhoods can also be used or vacant properties put aside for localised energy generation. Opportunities can also be found in privately-owned spaces. The UK is home to an estimated 250,000 hectares (2,500 km^2) of south-facing commercial roof space (BRE, 2016).[5] Placing solar panels on these roofs is not a niche policy option: in 2022, the UK Warehousing Association called for the change of energy models to allow warehouse owners to install rooftop solar. This call was based on the rising energy prices faced by these companies, However, while rooftop solar is often economically attractive, the roll-out of the technology on commercial roof space is, at times, held back by who owns it. Some 55 per cent of commercial property, by value, is rented rather than owned by the companies who occupy it (PIA, 2017). Future urban planning policies need to see these spaces as key parts of local renewable energy landscapes and as sites of investment and revenue generation and work with both occupiers and owners to use these spaces for energy generation.

Mass retrofitting now

Local and regional governments are well placed to work with the owners of buildings, like schools and warehouses, to install solar panels that benefit both the employer and the local community. However, national government action is also required to implement the wide-ranging schemes needed to address energy poverty. To date, government policies in the UK to help households be more energy-efficient have been either underwhelming or notable in their absence. Government schemes fell victim to the axe of the austerity economics adopted by the Conservative-led governments of 2010 onwards. Prime Minister David Cameron famously called for policy makers to 'cut the green crap' in 2013 – leading to subsidies for energy efficiency measures being gutted. The Zero-Carbon Homes Standard, that had aimed to make all new homes carbon neutral, was later scrapped in 2016.

Recent UK government policies have mainly targeted relatively low-hanging fruit, supporting single energy efficiency measures, rather than the broader transformation of a home (Putnam and Brown, 2021). The Green Homes Grant was introduced in 2020. It included £2 billion of funding to be distributed in small grants (up to £5,000, or £10,000 for those on lower incomes) to support making homes more energy-efficient. The scheme lasted less than a year, being quietly axed in 2021. Only 2,900 applicants (of a total of more than 120,000) received the required support to pay for energy-efficiency improvements (Welsh, 2021). The collapse of schemes like this isn't due to low demand or need. To date, only 49 per cent of houses in the UK have cavity or solid wall insulation and less than 40 per cent have loft insulation – both key energy efficiency measures (Nuttall, 2022). In recent years, the number of those installing such measures have slowed to a near halt (CCC, 2021). There is a lot of work to do to make the UK's homes warmer and more energy-efficient.

A central route to expanding rooftop solar in the UK is through adapting current housebuilding policies and regulations. Housebuilding is a key plank of economic policy in the UK and is seen to both address pressures on housing stock and provide economic stimulus in the wake of the COVID-19 pandemic. Before the COVID-19 pandemic, the housebuilding industry supported close to 700,000 jobs and was worth £38 billion (HBF, 2018). The pandemic slowed the sector – with housing supply dropping to a five-year low in 2021 (Heath, 2021). However, housebuilding groups already enjoy multi-dimensional support from the UK government that can be tailored and shifted to incorporate a greater focus on energy efficiency. UK government support to the sector comes in the form of development finance to small and medium house-builders and through providing infrastructural finance and support to larger developments and developers. Furthermore, the Land Release Fund (LRF) and One Public Estate (OPE) scheme provide funding to build new homes on publicly owned land that is deemed 'surplus' or 'unused'. There is also the Brownfield Land Release Fund that supports the redevelopment of former industrial or commercial land as new housing developments. Both schemes are closely tied to the Conservative government's 'levelling up' agenda, which aims to provide new forms of funding to help boost regional economies and create new employment opportunities.

There are government-defined regulations for new homes built in the UK that are, in theory at least, linked to climate action. From 2023, new-build homes must emit 31 per cent less CO2 than those in 2021, have better energy efficiency and be 'heat pump ready'. These regulations replace the previously-scrapped Net Zero Homes Standard, which had had envisioned households generating as much renewable energy as they demanded or, if this was not possible, making up for the shortfall through a renewable energy facility onsite. While the government in England scrapped this, the devolved government of Scotland expanded their house-building standards. The 2015 building regulations in Scotland provided a boost for rooftop solar – including a focus on the technology in its stipulations for energy demand and efficiency. It has been estimated that at least 60 per cent of all new homes built in Scotland in 2020 have rooftop solar panels (Hayes, nd.). The figure may even be closer to 80 per cent. Of the 61,455 UK properties that had rooftop solar installed in 2021, a quarter were in Scotland – which is home to just over 8 per cent of the total population of the UK (Reid, 2022).

Some housebuilding interests have previously lobbied against energy efficiency and decarbonisation requirements. The building company, Taylor Wimpey has previously lobbied against the Future Homes Standard that targeted the reduction of carbon emissions from homes built after 2024 by 75–80 per cent (Boren, 2021). This opposition can be based on costs. In 2022, the Home Builders Federation asserted that new regulations on housebuilding companies, including those linked to energy efficiency, added £4.5 billion to developer costs annually, representing more than £20,000 per new home (Hammond, 2022). The sector has been slow to embrace the changes needed to both decarbonise the housing sector and ensure energy efficiency. Between March 2021 and March 2022, close two-thirds of new homes built in England and Wales were built to use gas boilers for central heating, compared to 0.18 per cent which were built with heat pumps (ONS, 2022d). This is despite government policies banning gas boilers in new-build homes from 2025.

Any focus on restorative justice in energy transitions cannot only be on new-build homes alone. A national retrofitting programme should be a national priority. Energy demand strategies represent a significant route to addressing climate change, require less investment than more-technical solutions, and can be ratcheted up at scale and speed (Barrett et al, 2022). The 2022 report of the UK Committee on Climate Change highlighted a lack of home insulation and retrofitting policy in the UK as a particular failing of national climate action. Government policy must target the leakiest housing stock. The average UK home is rated as a D in energy performance certificate terms, halfway down the scale in terms of energy efficiency and running costs (A is the best, G the worst) (ONS, 2022c). Numerous local groups fill the gaps left by a lack of central government policy by providing time, expertise, and support to people seeking to reduce the energy that leaks from their homes. In Greater Manchester, the People Powered Retrofit scheme supports homeowners in both evaluating the effectiveness of energy efficiency measures and in finding contractors (CLES, 2020). In Bristol, the Cold Home Energy Efficiency Survey Experts scheme (CHEESE) provides direct assistance for homeowners looking to measure their current levels of energy efficiency through thermal imaging technology and expert guidance. Both schemes allow for the targeted action to reduce domestic heat losses and support residents in making small changes to keep their homes warmer and reduce their energy bills – and both on a not-for-profit model.

Despite energy inefficiency being present in all regions and communities, an energy crisis has hotspots. It's important to focus on where these inefficient homes are and who is living in them. Energy poverty has a particular geography, with many of the homes most-affected found in the Midlands and the North of England (ECIU, 2022). Of the 203 English constituencies found to have above-average rates of energy poverty, 55 per cent are found in the north of the country and a third in the Midlands (Gov. UK 2022c). These households are politically important: many of the areas that suffer the most from energy-inefficient homes are found in what were marginal constituencies in the 2019 general election (Grenville, 2022). It is also possible to identify which neighbourhoods should be prioritised in future energy efficiency schemes. An estimated 12 per cent of England's total carbon emissions are estimated to come from the 3.3 million homes built between 1918 and 1939 that have not been retrofitted – with some

17 per cent of households in these inter-war suburbs believed to be living in energy poverty (RIBA, 2022).

Future retrofitting policies should include a thermal prioritisation of needs (Brooks and Davoudi, 2014). For example, Glasgow is home to around 70,000 tenement flats. These buildings hold rich historical heritage and cultural value. With high ceilings, solid walls, and suspended timber flooring, they are often hard to heat – driving up energy bills and leading to cold homes. The percentage of residents in tenement flats in Scotland living in energy poverty might be as high as 30 per cent (175,000 households) (Fraser, 2021). There is a need to retrofit more of these properties, improving them and making them more liveable.

The subsidies required for a UK-wide retrofitting scheme might cost up to £3.6 billion of investment (EEIG, 2022). However, the net benefits of expansive energy efficiency measures in the UK could be worth over £55 billion, including benefits in terms of comfort, reduced emissions and better air quality and health (Eyre et al, 2017). With a prohibitive factor of people retrofitting their homes being the upfront cost, policies should offer financing tools that spread payments over longer timeframes. Inspiration for this can be found nearby. In the Republic of Ireland, the government has launched a National Retrofit Plan in 2022. This retrofit plan has dedicated €35 billion (£30 million) to energy efficiency measures. Measures included in this package addressed other elements of a just energy transition – including investments in supply chains and new jobs. It also incorporated elements of fairness (including a language of just transition) and universality (ensuring the benefits are available to all) (DECC Ireland, 2022). Key policies present in the plan in the Republic of Ireland included grants to support deep retrofit of buildings and the installation of cavity wall insulation, and free energy upgrades for those living in (or at risk of) energy poverty and/or those receiving certain welfare payments. This can be expanded even further. In the Republic of Ireland, some low-income households do not qualify for the free energy upgrades provided due to the arbitrary guidelines neglecting the fluidity and complexity of energy insecurity (McGlynn, 2022). With millions of households now at risk of energy poverty, government schemes need to be expanded and focus on providing help to improving the energy efficiency of all homes – to reduce emissions, decrease energy demand, and improve lives.

Rule 4: Prioritise those most vulnerable to energy poverty

The need for wholesale energy efficiency measures will not be news to anybody working in policy-making in the UK. The Conservative Party manifesto in the 2019 election pledged £9.2 billion of investment in greening homes and buildings and, in November 2022, the Chancellor of the Exchequer, Jeremy Hunt, announced a doubling of annual investment in energy efficiency measures. A few weeks later, the government of Prime Minister Rishi Sunak (2022–) announced £1 billion for energy saving measures, with eligible households able to request funds to cover 75 per cent of the cost for measures (Mavrokefalidis, 2022). However, Hunt's budget pledge in November 2022 came with a caveat: with any spending delayed for three years.

Moves toward energy efficiency fuse decarbonisation policies with aims of addressing fuel poverty and its various impacts, particularly for health. It also necessitates focusing on those most vulnerable to fuel poverty – targeting interventions to help people warm their homes. Future policies and schemes need to include restorative justice at their heart. The current energy model is unable to do so without a change of approach. With households across the UK facing increased bills and colder homes, there is an urgency for such a model to be altered and transformed. The policies that might be introduced to do so vary, ranging from the introduction of new social price tariffs to the nationalisation of energy companies (Robinson and Simcock, 2022).

In an era of rising prices and 'disconnections by the back door', there is a need to affirm a right to heat and light in our homes. Rooftop solar and household retrofitting intersect with numerous policy objectives – decarbonisation, energy poverty, energy justice, and green jobs and skills. Retrofitting programmes improve the energy efficiency of people's homes and make them nicer, more comfortable places to live. Filling holes in walls not only stops draughts and heat escape – it also helps reduce indoor mould and the negative health impacts that come with it. A claim to 'Universal Basic Energy' would necessitate a restructuring of the current pricing model to assume that every household is guarantee a certain amount of energy for free (Robinson and Simcock, 2022). This amount would include energy for cooking food, heating our homes, and lighting our evenings. Different people and households, such as those with disabilities or medical problems, would be prioritised and benefit from larger allowances to make space for the electricity that they need to use. This approach could be underpinned by the expansion of rooftop and locally sited solar to directly provide electricity to those living nearby. This would not only allow for more people to use cheaper electricity but would also lessen pressures on national energy grids: with such technologies helping to reduce moments of peak demand.

Solar panels should be on every rooftop. They should be desirable selling points of a home for sale or a new rental, like a newly fitted bathroom or marble countertops.[6] They must become commonplace, found on the roofs of homes, commercial buildings, and public spaces. A landmark scheme that targets the expansive roll-out of rooftop solar to homes and publicly owned buildings is the '30 Million Homes' campaign in the United States.

This campaign, led by a national coalition over 220 organisations, called for enough rooftop and community-owned solar to power 30 million residences across the country in the next five years. The campaign lobbied for the expansion of a current scheme of solar tax credits to include a cash grant, allowing more people to access and use it. It is estimated that, if implemented, This scheme would benefit one in four households in the country, and lower total energy bills by an estimated $20 billion (Solar United Neighbours et al, 2020).

New forms of funding that enable and empower more people and households to install rooftop solar or energy efficiency measures include zero interest loans, low interest finance, or grants targeted at those on lower incomes (Brown and Bailey, 2022). Residents could also lease solar panels, rather than buying them outright, with a third party owning the system but the electricity provided going to the property itself. Millions of households enter in similar arrangements of high value when buying a car, with monthly payments, in essence, paying for a long-term rental. This may go some way to reduce the up-front costs that can often stop installation (O'Shaughnessy et al, 2020). Financing could also be expanded and made more equitable by introducing a 'Pay as you Save' form of tariff on electricity bills, allowing for post-installation cost savings to pay-back any loans or lease payments due.

Local authorities and councils are exploring more and more local energy efficiency programmes in their housing stock. It's easy to see why: with many declaring climate emergencies and calling for climate action, the housing stock that they own and operate represent places where a significant proportion of their organisational emissions might be saved.[7] Where councils don't own social housing stock, it is held by companies who hold expansive influence over local energy demand and emissions, with social housing providers owning whole streets and apartment blocks. Retrofitting and the rollout of rooftop solar on social housing can be transformative for the residents living there, who are often those on the lowest incomes and the most vulnerable to energy price rises. Such policies will also allow greater recognition of how people use energy differently. For example, older residents may have different patterns of household demand (Willand and Horne, 2018). In having both knowledge and engagement with their residents, local authorities and housing associations may be able to tailor approaches to residents and their needs.

Investing in the retrofitting of social housing would be restorative: providing healthier and better living conditions for vulnerable households and ensuring that the benefits of energy efficiency measures are shared equitably. Between 2016 and 2018, around 3.9 million people in England lived in social housing, renting their homes from a local council or housing association (Gov.uk, 2020). This is skewed along lines of race and ethnicity – with households defined as 'Black African', 'Black Caribbean' or 'Mixed White and Black

African' most likely to rent social housing. Broadly, households understood as Black and Minority Ethnic (BAME) in the UK are over twice as likely to live in persistent poverty than White families (Social Metrics Commission, 2020).

Local governments can support local jobs and companies through retrofitting social housing, encouraging new developments to include measures like solar panels and heat pumps, and supporting further education programmes that support those seeking to learn these skills. This is not a new claim. The German politician and architect of *Energiewende* policies, Hermann Scheer (2001) argued that locally owned utilities can obtain the rights to lease and use the roofs of private buildings, creating local networks of solar panels. Locally driven retrofitting work can also be linked to the community wealth building processes discussed in Chapter 5. Energy efficiency schemes can provide business opportunities for local businesses, creating new revenue streams and stimulating new jobs in the local economy. The Centre for Local Economic Strategies (2022) in the UK has highlighted how local construction firms should be targeted by policies that provide training opportunities and support them in expanding or pivoting their businesses to include retrofitting work.

We need people to complete this work. We need workers to fit solar panels, insulate roofs, and retrofit houses. While often labelled as resistant to decarbonisation, workers can share the values and mission of community energy ventures, in terms of having their voices heard and protecting communities. Energy transitions will bring with them a process of restructuring which will change patterns of employment for many. In the UK, a nationwide energy efficiency retrofitting programme could create over 400,000 jobs (Krebel et al, 2020). These jobs will be highly important in regions where jobs are at risk from decarbonisation, such as in the Yorkshire and Humber region where up to 360,000 jobs are currently based in carbon-heavy industries (Diski, 2021a). This, in turn, can go some way to recognising and addressing distributive justices, particularly in contexts of austerity and rising prices. Somebody needs to do this work and they need to have a voice in what decarbonisation might look like.

7

Work

The road to Blyth runs along the coast, stretching from Whitley Bay to the Wansbeck Estuary of the River Blyth in the northeast of England. Numerous landmarks hint at the region's past. Remnants of Second World War batteries and searchlight sites look out to a lone fishing boat pulling into the harbour. The port of Blyth dates to the 1100s when it was used by a nearby monastery to ship salt. The town prospered from the 18th century onwards through coal mining and shipbuilding, with the expansion of the railroads allowing Blyth to become a key port for the export of coal to the European continent. This history is memorialised today. Standing on the quayside is the *Spirit of the Staithes*, a sculpture by Simon Packard that memorialises the spaces where coal was stored before being transported elsewhere.

During the miners' strike of the 1980s, striking workers would come to the quayside where this sculpture stands to look for sea coal, while others would fish or work on their allotments (Samuel et al, 1986). Today, men fish off the quayside, near buildings with plaques memorialising industries long gone. The Blyth Shipbuilding Company (opened in 1883) closed in 1966 at the cost of almost 1,000 jobs (Milne, 1966). This was, to some extent, eased by the construction of the coal-fired Blyth Power Station (in 1958) and, later, the Alcan Lynemouth Aluminium Smelter (1974). Yet, these sites of work closed within two generations – being shut down in 2001 and 2012 respectively. Rather than locked into fossil fuel energy infrastructure, Blyth found itself cut adrift.

Today, dog walkers and couples on the beachfront eat ice cream and cast their eyes at five wind turbines in the mid-distance. These turbines power 36,000 homes and are symbols of the entrance of renewable energies into Blyth's future. The Blyth Harbour Wind Farm was initially commissioned in 1993, with an offshore facility (the first of its kind in the UK) built in 2000 and later replaced by these five turbines in 2019. This coastal town, a short journey from Newcastle, has become a key site of national energy transitions in the UK. Today, a walk around Blyth is a series of steps around key sites and infrastructures of the UK's decarbonisation ambitions. The old shipyard now houses the National Renewable Energy Centre, set up in 2002 to develop and test new technologies. The South Harbour Terminal of the Port of Blyth houses heavy goods and large cranes, mechanical trenchers, and pipeline and cable ploughs. Located almost halfway between Great Yarmouth in Norfolk and Aberdeen, Scotland, the facility is a key site for the expansion of offshore wind and the decommissioning of oil and gas platforms in the North Sea.

The buzz around the South Harbour Terminal corresponds to a concerted strategy, adopted by the Blyth Port Authority, that has placed renewable energies at the centre of regional economic change. In 2021, the Port of Blyth Authority announced the new Bates Clean Energy Terminal, providing new facilities for companies working in renewables and offshore (Whitefield, 2021).[1] The Authority's position as an independent statutory trust (awarded by the UK's Houses of Parliament in 1882) means that it doesn't have private shareholders or owners and, instead, is controlled by an independent local board. This has allowed it to reinvent itself dramatically in response to the decline of other activities at the Port. Its ownership model as a public trust allows it to take risks – foregrounding decisions in social benefits rather than the immediate financial priorities that restrict the actions of other ports, which are primarily privately owned in the UK (Evans, 2020). The Port of Blyth invests profits in facilities and community outreach. It offers training and skills development, partners with universities and colleges, and supports community organisations and events.

Across the Wansbeck Estuary from Blyth lies more activity, including the construction of the Northumberland Energy Park, a deep-water base for the offshore energy industry that represents close to £35 million in investment in the regional economy. Further inland at East Sleekburn, lies the great North Sea Link that connects the UK national grid to up to 1,400 MW generated by Norwegian hydropower, the longest deep-sea interconnector in the world (at 720 km long). The Port of Blyth is unrecognisable to how it appeared two decades ago. Its reinvention demonstrates how the renewables sector represents more than mere technological change and new infrastructure. It also offers new opportunities for skills, jobs, and brighter futures for those living there. Perhaps you might find a house and family here that has directly experienced this change across its generations – with a grandfather employed in coal mining, a son at the power plant, and a granddaughter soon to be employed in the offshore energy industry.[2]

It is also in Blyth that national decarbonisation ambitions have stalled. The 235-acre site of the old coal power station, just outside of the small coastal village of Cambois, was due to be a key site in the UK's energy transitions and the greatest investment this region had seen in generations (BBC, 2020). The Britishvolt factory was a planned £3.8 billion factory that would manufacture the batteries used in electric vehicles. The facility promised

3,000 highly skilled jobs, training opportunities, and a new future for many. With 28,000 jobs in the north of England deemed at risk in decarbonisation agendas, this scheme was of great importance to many (Halliday, 2018). As one comment submitted in the planning permission process put it: 'I am a Teacher in Blyth and see first-hand what high unemployment can do to communities and would very much like to see more future for the children of the North. So many have to leave the area because there is so little modern economic opportunity.'[3]

However, the Britishvolt investment in Blyth stalled in the autumn of 2022. While the company was celebrated as the future of the regional economy by politicians, it struggled to raise the funds necessary to build its Blyth facility. The national government had pledged £100 million of investment but this funding was linked to milestones in project construction. When these were missed, the government refused a request to provide emergency funding. Private investment was, in turn, linked to this government payment – meaning that, Britishvolt was unable to secure the money needed to start working towards those agreed milestones. In October 2022, it was reported that the company, previously accused of chaotic management and an unclear business model, only had one month's worth of funding remaining (Campbell et al, 2022a; 2022b). Workers took a pay cut in November 2022 in an attempt to save the company from bankruptcy (Sillars, 2022). Britishvolt, seen as the great hope of the UK's electric vehicle industry only years before, faced ruin.

An energy transition will have important impacts on local, national, and global economies. New jobs will be created by the expansion of low-carbon products and infrastructure; others will replace similar but more polluting roles; carbon-heavy jobs may be phased out completely. A failure to recognise and account for these changes to employment will give rise to new forms of injustice, with significant knock-on impacts. Just as there is much talk of financial assets becoming stranded by climate change and decarbonisation (Livsey, 2020), there are communities across the globe at risk of being left stranded by a decarbonised economy. In the UK, stranded communities might come to exist in places like Aberdeen and Grangemouth, which are economically reliant upon oil and gas. Other communities at risk are not necessarily linked to fossil fuel extraction but include the steel, glass, and transport industries. Decarbonisation may come to represent the terminal and prescribed decline of these places, in cultural as well as economic terms.[4] The historic ties between these communities and carbon-heavy work are often set deep, with fossil fuel extraction and technologies interwoven with cultural identity and everyday life (Huber, 2015; Mayer, 2018). For many living in stranded communities, decarbonisation can come to be perceived as a political decision and as a punishment (Olson-Hazoun, 2018). This suspicion is evident in the words of Gary Smith, the General Secretary of the GMB trade union in the UK:

> [We] should stop pretending that we're in alliance with them [environmentalists]. The big winners from renewables have been the wealthy and big corporate interests. Invariably the only jobs that are created when wind farms get put up, particularly onshore wind, have been jobs in public relations and jobs for lawyers. (Wearmouth, 2022)

For Smith, energy transitions represent more of the same, with private companies and the professional classes benefiting, while the members of his union bear the burden of lost jobs. Across documents that detail just energy transition approaches, the protection of jobs and security of livelihoods is a central element. This is also evident in national climate policies, which promise millions of new jobs: for example, the UK government has pledged to create 2 million new green jobs by 2030 (BEIS, 2020b). These policies have broader consequences: providing new forms of employment, changing local and regional economies, and supporting new generations. If a call for a 'just energy transition' pushes back against a technical understanding of decarbonisation, the development of new plans and policies to create and protect jobs provides an important route to supporting those communities who anticipate, fear, or resist the impacts of energy transitions. There is a need to put workers – and work itself – at the centre of any just energy transition.

Green jobs

Statistics released by the International Energy Agency in 2022 suggested that 'green' energy jobs now outnumber fossil fuel jobs in nearly every region of the world (IEA, 2022c). In total 12.7 million people were employed in the renewable energy sector in 2021, a figure nearly twice as large as it was ten years previously (Ferris, 2022). However, gauging the success of green jobs schemes is challenged by a lack of a clear definition of what such a job looks like. The types of jobs included in the International Energy Agency's numbers are highly varied, requiring different skills and levels of education, and leading to varying levels of pay, job security and working conditions. Hundreds of unique occupations in the energy sector may qualify as 'green' — ranging from electricians and powerplant operators to those tasked with hazardous waste removal or retrofitting houses with energy efficiency measures (Muro et al, 2019). For clarity, I follow the definition of 'green jobs' provided by the International Trade Union Congress:

> A green job reduces the environmental impacts of enterprises and economic sectors to sustainable levels, while providing decent work

and living conditions to all those involved in production and ensures workers' rights are respected. Green jobs are not only those traditional jobs people think of as green – like making solar panels, manufacturing wind turbines, water conservation and sustainable forestry. They also include retrofitting related jobs in the construction and public transport sectors, and making energy efficiency improvements in manufacturing plants, along with services supporting all industries. (International Trade Union Congress, 2012: 3)

This definition broadens what 'green' work is to include the numerous possibilities of a low-carbon economy. It is important to expand notions of decarbonised work beyond the construction, manufacturing and energy sectors that are at the centre of current policies (Stevis, 2018b). While many focus on which jobs can be 'greened', it is also necessary to think about what jobs can be *improved* for the workers themselves. Good, green work should not be limited to certain sectors but available to all. This involves broadening our understanding of green jobs to include those roles that may not, at first, seem green, but represent work that already has low emissions and supports people and communities. This might include supply chains but also other roles. A just energy transition shouldn't just be for those who fit heat pumps or build wind turbines. It should also be for others: like hospital staff, social care workers, and cleaners, who play an important (if overlooked) social role in supporting climate action and adapting to its impacts.[5]

The International Trade Union Congress's definition of green jobs is also useful in its adoption of a language of 'decent' work. This allows for the critical reflection on the extent to which contemporary green jobs are 'good' jobs. The International Labor Organization (2021) have defined 'decent' work as incorporating:

- job creation (generating new opportunities and alternatives to current or past employment);
- the fulfilment of rights at work (ensuring employment rights are sustained, protected and expanded as necessary);
- social protections (allowing for a broader safety net for those underemployed or unemployed and the development of skills training); and
- social dialogue (the creation of channels for companies, trade unions and governments to define future continuity and change).

A call for 'decent' work within a just energy transition asserts that decarbonisation policies can play an active part in improving people's lives, not just providing a basic level of employment. Embracing the role of decent work means that any move to a 'green' job should not lead to any decline in working conditions. In linking this to 'social dialogue', the definition of decent work asserts that workers and employers cease to be mere bystanders in decarbonisation. Instead, they can take key roles in the creation of new, better forms of employment.

Decarbonisation will involve a process of economic restructuring, in which local economies, jobs and communities will be forced to change. Those negatively impacted by decarbonisation not only include the fossil fuel companies and their executives, whose wealth is based on extraction and high emissions. It will also include recognising the communities who have had their livelihoods and local economies based around emitting industries and how a just energy transition can replace these jobs with better ones. Fossil fuel industries produce a boom for local employment: not only at an oil refinery or coal mine itself but often in the subsidiary industries that grow to support these central workplaces. In many places, there can be almost as many jobs in these subsidiary industries as the original fossil fuel jobs themselves (Marchand, 2012). All of these roles are now at risk in decarbonisation. The closure of the Hazelwood mine in the Latrobe Valley, Australia resulted in the loss of 1,400 direct jobs, an impacted worsened by the associated decline in spending in the area (due to lost wages) leading to a further 1,771 job losses (Sheldon et al, 2018). Many communities are 'locked in' and dependent upon carbon-heavy work and the fossil fuel economy and, as a result, are at risk of being left stranded in decarbonisation agendas.

If climate targets to reduce global temperature change by 2°C by 2050 are met, there will be an estimated 26 million jobs globally in the energy sector alone, with renewable energy jobs both acting to replace lost fossil fuel jobs and create additional roles and opportunities (Ferroukhi et al, 2020; Pai et al, 2021). It has been estimated that the growing solar industry in the United States can provide a new home for all coal workers at risk of unemployment over 15 years, and that the potential for wind energy jobs broadly correlates with the numbers of those at risk of fossil fuel unemployment (Louie and Pearce, 2016; Tomer et al, 2021). In many communities, the arrival of renewable energy jobs can represent new economic beginnings – bringing new jobs and revenues to local communities, businesses, and governments and filling the gaps left by declining industries (Martin, 2021). In Northern Germany, the offshore wind industry provided new workplaces for those made redundant by declining shipyards, who now build bases and towers for offshore infrastructure (Fornahl et al, 2012). This can be transformative. The arrival of new renewable energy manufacturers in the rural of Dobbe, also in Germany, created new forms of employment that allowed younger generations to stay in the region, slowing a population exodus that had characterised the area previously (Carlson, 2019). In Poland and Belgium, an estimated 15 per cent of young people now find their first job in the green jobs sector (Sulich et al, 2020).

The promise of jobs is significant in any renewable energy transition – providing an important route to building a coalition of different groups and communities. A key focus of a just energy transition is on how policies must be enacted to trace the jobs lost in decarbonisation and mitigate the impacts that such declines may have on individuals, households, and communities – across the energy sector. It's important to reflect on what these new jobs in the renewable energy sector might look like – and who they might be accessible to.

Good jobs?

For an energy transition to be socially just, green jobs must be *better* than the ones that are being lost in a future low-carbon energy grid. In an era of financial pressures and precarious work, new jobs must promise a brighter future, not more of the same. With green jobs giving a reason for younger people to stay in regions and build their lives in certain places and support decarbonisation through their labour, those lives must have promise, they must be better than the alternatives in any continuation of the status quo. Similarly, those workers who are witnessing their carbon-heavy roles become redundant must see a secure pathway for them to transition to new job roles that are better than the ones they had before. A failure to provide these will lead to injustice in energy transitions, with workers forced to bear the negative impacts of decarbonisation. This will further increase the suspicion of many towards energy transitions.

In the current energy model, promises of 'good, green jobs' have not fully materialised. The wages offered by the renewables sector still rank far below the average annual pay of those working in oil and gas, the exact sectors that society is looking to transition from (Global Energy Talent Index, 2021). In New Jersey, US, the median wage paid to a solar installer is $43,620 compared to the $80,530 median annual salary of those working in fossil fuel power plants (Vachon, 2019). Workers on solar projects in the United States can earn a third of what they might have done in equivalent fossil fuel jobs previously (Scheiber, 2021). In the UK, the renewables sector has been found in several cases to be responsible for dramatically low levels of pay. Ships installing offshore wind facilities off the UK coast have been found paying far below the state-mandated National Minimum Wage (£9.50 per hour in April 2022). Many workers, often international migrants to the region and industry, sign fixed-term contracts at a certain wage, that is not altered to consider hours worked overtime – driving down hourly rates.[6] In 2021, a ship called the *Glomar Wave*, operating from Great Yarmouth was accused of be paying a daily rate of £47.35 which, when broken down by the 12 hours worked by the crew, equalled £3.94 per hour – less than half the National Minimum Wage (£8.91 in 2021) (RMT, 2021). In Grimsby, catering staff on board the *Edda Passat*, which was servicing a wind farm off the Lincolnshire coast, workers were reportedly paid £6.75 an hour, again below minimum wage (Daly, 2019). In Scotland, Russian and Indonesian workers have been found to be being paid less than £5 an hour for work offshore (Lawrence and McSweeney, 2018; STUC, 2019). The workers worked 12-hour days, seven days a week for the extent of their contract.

Renewable energy jobs also involve different working conditions. While a transition away from the occupational health risks and hazards associated with coal mining and other polluting industries should be celebrated, new workplace hazards have emerged. The manufacturing process of solar PV technologies exposes workers to hazardous chemicals such as arsenic, lead and cadmium (Spinazzè et al, 2015). Working from height at wind farms has led to deaths – including that of Antonio Joao da Silva Linares at the Kilgallioch site (BBC, 2017) and an unnamed Spanish man at Whitelee Windfarm within weeks of each other in Scotland in 2017 (Ward, 2017). In 2018, Ronnie Alexander, a security guard died at the Afton wind farm, also in Scotland, freezing to death after being caught in a snowstorm. His employers admitted breaches of health and safety rules led to his death (Rutherford, 2021). They were fined £900,000 in November 2021 (BBC, 2021). While the manufacturing and construction of renewable energy infrastructure might promise new jobs and increase employment in communities, these processes may also lead to new health impacts that can intersect with broader, entrenched patterns of inequality.

Renewable energy work is illustrative of energy transitions being understood within a relatively short-term lens, with the primary goal being emissions reductions, rather than a focus on securing long-term employment on good terms. This results in a precarity of the work itself. The people decarbonising the energy grid are not always well-paid, nor do they always enjoy the benefits of long-term and secure employment. In the US, many workers installing solar arrays are employed on temporary contracts and move from project to project and region to region (Gurley, 2022). The temporary nature of this work also leads to an increased complexity of negotiating pay for individual projects, rather than having secure jobs and wages. This is part of a broader process of increased casualisation and precarity of work which is being repeated in the emergent renewable energy sector. The construction of a wind farm requires a burst of labour-power to manufacture parts and assemble them into a turbine. The solar industry operates as a moving assembly line: with workers performing specific tasks on each panel across the site. Once this work is done, the labour requirements shrink dramatically to those required for maintenance only.

For many, the transition to green jobs not only represents a change in employment but also a departure from the gains made by the organised labour movement itself. Many fossil fuel jobs are unionised and, as a result, involve strong worker protections. These have come as the results of decades of struggle by generations of workers (Vachon, 2019). Yet, these jobs are in decline and will be replaced by others that might fail to inspire much optimism – being lower-skilled, non-unionised, and precarious (Sheldon et al, 2018). Only 10 per cent of solar workers in the United States are unionised, restricting the bargaining power that those employed in the industry hold (IREC, 2021). These jobs don't necessarily promise a new future, nor do they fulfil the promise of a new industrial revolution that creates new livelihoods and wealth. These workers remain vulnerable to precarity, poor pay and poor working conditions in the current model of energy transitions.

Renewable jobs might be some of the fastest growing jobs but they are not available to all workers. Green jobs are more likely to be held by White men, aged between 35 and 64 and with a high school education (McClure et al, 2017). In the United States, Black workers hold less than 8 per cent of the jobs in the sector, despite representing 12.4 per cent of the total population (US Department of Energy, 2022; Jones et al, 2021). Latinx workers make up around 17 per cent but primarily work in entry-level, low-income installation work, jobs roles paid far less than colleagues in other parts of the industry (Jones, 2021; IREC, 2021). These Black and Latinx workers have historically experienced less stable employment and suffer a bigger threat of redundancy during economic downturns (Reid and Rubin, 2003; Penner, 2008). The lack of employment opportunities represents yet another stage in a series of environmental and energy injustices explored in previous chapters, which includes fossil fuel energy projects being built on their doorsteps and a lack of access to the new renewable energy technologies themselves.

The accessibility of green jobs is also skewed in terms of gender. In 2019, 32 per cent of the renewable energy workforce were women — compared to a 22 per cent share in the oil and gas industry (IRENA, 2019). However, this statistic conceals that many women were employed in primarily administrative roles (Clancy and Feenstra, 2019). Within the sector, 45 per cent of administrative roles in the sector in 2019 were held by women, compared to 28 per cent in roles defined as specialist 'STEM jobs' (linked to 'Science, technology, engineering, and mathematics') (IRENA, 2019). There is also a lack of gender representation in positions of management and leadership in the renewables sector, with women occupying only 18 per cent of management positions (C3E International, 2019). The causes for this underrepresentation are structural (IRENA, 2020a; 2021c). Working conditions in the sector can often represent spaces where women are made to feel unwelcome: with issues including comments about appearance, sexualised comments, complaints about their presence on-site, or being made to feel like 'token women'.[7] Outdated styles of leadership, unconscious bias, and a lack of role models in the industry have also been cited as key reasons for women leaving jobs the renewables sector (GWNET, 2019).

The renewable energy transition is creating new jobs, but these opportunities exist along the old and outdated lines of gendered exclusion and inequality. This has important consequences. Decarbonisation and broader energy decision-making are often presented in overly technical terms that fail to recognise, understand, or address gendered inequalities in who has access to and control over energy (Mang-Benza, 2021). This represents a procedural injustice, with the continued absence of women in energy policy excluding them from decision-making processes. Better representation within the sector would allow for a diversity of perspectives that allow for broader, more inclusive approaches to energy transitions.[8] Furthermore, women have been found to often be more likely to support pro-environmental behaviour and renewable energy transitions, meaning that efforts to increase gender diversity might also allow for the influx of employees who already support the mission itself and are able to speak more directly to those interested in adopting renewable energy (Allison et al, 2019). In doing so, they can dictate a move beyond the gender 'blindness' that fails to address contemporary issues in energy politics and policy and recognise patterns of inequality (Lieu et al, 2020).

The jobs created by decarbonisation are precarious, hazardous, inaccessible, and paid less than the carbon-heavy jobs that policy makers are encouraging people to leave and transition from. Aims for rapid decarbonisation and hopes for new jobs and livelihoods don't match up. The construction of renewable energy infrastructure requires labour that is short-term, precarious, dangerous, and not accessible to all. This is rooted in the urgency of energy transition, with short-term goals of the expansion of renewables leading to policy makers and employers overlooking the need for new 'green' jobs to be as good as the jobs that are being lost. One solution to this is a refocus energy transitions on green skills, rather than green jobs only. This move would ensure that workers in an energy transition can develop the skills that allow them to be in demand across the energy sector, and new training schemes include a focus on ensuring better representation and equality in who gets these new jobs.

Skills and retraining in transitions

Contemporary energy transitions are challenged by a 'chicken and egg' type problem: what comes first, green jobs themselves or the green skills and training that they will require? (Cha, 2020). Government policies on green jobs often seem to tread the line of compromise between the two. However, this creates issues. It is difficult to create demand for technologies like rooftop solar and heat pumps if there is no labour available to install them. Waiting lists lengthen, prices increase, and more and more people see these technologies as something beyond their reach. The green skills required are varied and multi-dimensional. They might include the ability to work at height (helpful for work on wind turbines), electrical work (installing rooftop solar) or mechanical skill (fixing electric cars). While many skills are transferable from one sector to another, the need for reskilling and retraining is significant. In the UK, an estimated 6.3 million roles (or one in five jobs) will likely experience some change to the skills required (UK Green Jobs Taskforce, 2021).

A gap between the skills needed and the training available to workers is already providing a major barrier to decarbonisation in the UK, with it proving difficult to find the workers needed to enable the comprehensive shifts in energy generation and use required. This challenge emerges from several issues, including an underappreciation of the pace at which certain labour demands might grow; a general shortage of skilled workers in sectors, the poor reputation of certain sectors amongst workers, and a shortage of those required to provide the skills-based training required (Strietska-Ilina et al, 2011). In the UK, there may be as many as 41,000 jobs without workers with the skills required by 2030 (Keating, 2022).

To address this skills gap, all future energy policies need to include a focus on education, training, and skills. The UK Green Jobs Taskforce (2021), set up by the government in 2020, called for the embedding of green skills and knowledge into the curricula of both school- and college-level education. A focus on skills allows for the proactive training of young people and new entrants into the job market, providing them with long-term skills that support them for years to come. Apprenticeships are a key route to do this. In the UK, Friends of the Earth (2021) have called for a vast expansion of current apprenticeship schemes to create 250,000 'green' apprenticeships and provide young people, a group that has been disproportionately impacted by the COVID-19 pandemic and recent financial crises, with a direct route into employment when others might have been closed off by economic downturns.

Funding and support are needed for schools and colleges to host programmes that support workers seeking new green jobs and to appoint the staff required to train new students. Targeted curricula reforms have created programmes that can both provide training and a defined pathway to future green employment. This often requires the development of partnerships between colleges and renewable energy companies to ensure that education and training meet the requirements of the latter. For example, the wind energy company, Ørsted has worked with the Grimsby Institute in Lincolnshire and Furness College, Cumbria to run three-year apprenticeships in Maintenance and Operations Engineering (Friends of the Earth, 2021). In Navarre, Spain, the *Centro de Referencia Nacional en Energías*

Renovables y Eficiencia Energética has been created to support workers moving into the renewables sector and support a dramatic expansion of renewable infrastructure in the province (UNFCCC, 2020). Elsewhere, cooperatives and community wealth-building schemes have placed renewable energy at the centre of local skills programmes. The Evergreen Cooperative in Cleveland, Ohio in the United States (a precursor to the Preston Model discussed in Chapter 5) channelled funding towards energy solutions projects, in which workers were trained to install solar panels and other energy efficiency measures. In doing so, the cooperative became a regional leader in installing new energy technologies and provided important employment and training opportunities to workers in the region.

Opportunities for retraining also need to be provided for those who face losing their current jobs in fossil fuel and carbon-heavy industries. This is the core of a just transition approach: with previous policies to manage the restructuring caused by the closure of coal mines providing training schemes to help workers find new jobs (Galgósczi, 2014; Snell, 2018; Cha, 2020). Many of the skills needed for work in the fossil fuel sector are also required in renewable energy jobs (STUC and Transition Economics, 2021). A prominent example is found in how those working in the oil and gas sector hold skills necessary for new roles in offshore wind, such as working from height. It has been estimated that over 90 per cent of those working in fossil fuels in the UK have high skills transferability to green jobs, with an estimated 100,000 roles in the renewables sectors projected to be taken by those workers joining the sector from offshore oil and gas (UK Green Jobs Taskforce, 2021).

However, the way that systems of training and skills development are set up in the UK has blocked workers' skills transfer and transition. Training schemes are often individually accredited and, while they teach similar skills, there is a limited transferability of accreditation of workers having completed a course in one sector to the other. This can often lead to workers taking the same training programme twice, albeit for different sectors, and having to pay for both. It is hard to escape the sense that these companies are making money off workers who are trying to reskill.[9] The only way this can change is through collaborative action by the two sectors or new policies from the government. One way to do this is through the introduction of a skills 'passport', which oil and gas industry workers can use to evidence that they have the skills required to work in the wind industry. This would make the transition from one sector to the other, be it on a permanent or more-temporary basis, easier and support workers in securing income across the two industries. It would also allow the unlocking of the skills required for the expansion of renewable energy transitions and the alignment of training and standards frameworks of one sector with the other.

For individual workers, any transition won't be easy. Many of those facing reskilling have been in fossil fuel work for decades, have families, and need secure incomes. As a result, they may not be particularly welcoming of calls for them to return to the classroom to learn new skills. They require support and the opportunity to move to another sector as easily as possible. Education and training programmes can be modularised so that those on lower incomes or with time constraints can drop in and out or combine education with remaining in work, with financial assistance programmes also able to support their continued, if staggered, enrolment. Programmes can also focus to ensure diversity in the sector in the future. There are several schemes that illustrate this in the United States. In Newark, New Jersey, the Garden State Alliance for a New Economy has provided a short, six-week programme in green construction that focuses on training lower-income workers in new skills (Martinson et al, 2010). Educational, skills and training programmes can also be focused on ensuring gender parity in the pipeline to renewables jobs.[10] The gender gap in the renewables sector is itself perpetuated by a gendered way of seeing STEM skills, work, and training (Lieu et al, 2020). Any green jobs, skills and training programme must incorporate plans to address education inequalities and diversify STEM education (Kwak, 2021).

An important example of the targeting of training opportunities can be found in the GRID Alternatives project in Oakland, California. This targeted training programmes to improve diversity and inclusion in the solar energy sector, including targeting women, Indigenous groups, and those recently released from prison (Stephens, 2021). Its training programmes are based on not only skills provision but also increasing the employment opportunities for those enrolled, connecting them to industry, and creating supportive working environments to ensure employee retention (Stephens, 2021). More than 30,000 people have participated in these schemes and gained new skills and opportunities. There are similar schemes elsewhere in the US. In Louisiana, the Louisiana Green Corps has trained young unemployed people in energy efficiency retrofitting work, both improving local homes and providing new opportunities (Finley-Brook and Holloman, 2016).

Also in the United States, the nationwide Solar Ready Vets Network supports veterans or other military service members in gaining the skills, experience and contacts to join the solar energy workforce. This includes providing opportunities for on-the-job training in solar energy, as part of the wider support provided by the United States government to help those leaving military service find new work. Initially launched as a pilot in 2014, Solar Ready Vets has expanded and now seen over 500 veterans graduate from its programme. As of 2020, military veterans made up 9 per cent of the solar workforce (IREC, 2021). This pipeline can be transformative for many veterans, who face numerous challenges when leaving the military, such as a struggle to find work (O'Donnell and Rinaldi, 2022). A roll-out of a similar scheme in the UK could provide a route to new jobs for those who have left the armed services, particularly those who are vulnerable and require further support. In the UK, veterans can be at risk of homelessness. In Glasgow in Scotland, a survey of the local homeless population found 12% had previously served in the Armed Forces (Royal British Legion, 2017). They may also face issues of post-traumatic stress disorder, addiction, or mental health disorders (Olenick et al, 2015).

There is an important parallel for contemporary green skills policies in the UK – highlighting the need for a long-term approach to green skills and employment. In 2022, E3G, the climate change think-tank called for an 'Olympics-style skills and training programme' to train the workers required for the work required for decarbonisation (Phillips et al, 2022). The 2012 London Olympics promised a legacy of job creation and training for thousands of residents, including through language lessons and other programmes for those volunteering at the Games. At the time, the event was reported to have created 100,000 new jobs in London in 2012 (Prynn, 2012). However, many of these were part-time or temporary roles. While the International Olympic Committee (2021) argued nine years later that 110,000 more jobs had been created since 2012, local media has reported that many other promised jobs have failed to materialise, with the Queen Elizabeth Olympic Park employing far fewer people and in fewer workplaces than originally pledged to the community (Bartholomew, 2018). Large projects and schemes have promised new jobs and skills before, only for them to later disappear. The 2012 London Olympics employment programmes show what can happen when government policy focuses on a burst of skills, training, and job creation. However, it also highlights the importance of a just energy transition including a proactive, and comprehensive approach to green skills and employment that extends beyond the short-term construction of infrastructure and incorporates a broader vision of good work.

A just energy transition should focus on the development of skills – both to bridge any 'skills gap' hindering a transition and to provide generations of workers with new lines of work and secure livelihoods. To ensure that green jobs are 'good', such an approach must shift its focus from the immediacy of decarbonisation (and the construction of renewable energy projects only) towards a broader understanding of the complex networks of supply chains and subsidiary industries that support such a process. It is at these sites that workers present their own vision of a just energy transition – and how they fit within it.

Empowering workers across supply chains

Decarbonisation can happen without the jobs that are often promised to accompany it. For example, in Scotland, it has been estimated that decarbonisation could create up to 70,000 jobs or perhaps just 9,000 (STUC and Transition Economics, 2021). This is a vast difference and the route taken depends on the policies implemented by governments. Green jobs in the renewables sector can be low paid when compared to fossil fuel work, precarious and inaccessible to many. Employment in the renewable energy sector can often be concentrated at the construction sites for new infrastructure. This introduces a short-term character to the work: once a project is completed, the number of workers it employs drops to those who maintain it only. This has important consequences for local economies, that witness declining economic benefits and employment numbers after a project has been built (Pearse and Bryant, 2022).

The short-term logic of decarbonisation is limiting the potential of green jobs to be good jobs. A just energy transition will be built by workers – and it must be designed with them in mind too (Huber, 2022). The lack of representation, low pay and poor conditions presented above highlight the importance of ensuring that the voices of workers and trade unions are heard in renewable energy transitions. Green jobs represent the key ground upon which workers (and trade unions) and communities engage with climate action and seek to influence policy. While there is often an assumed contest between environmental protection and jobs that leads workers' groups to challenge environmental policies and decarbonisation, the labour movement has a rich history of pro-environmental action, both in terms of protecting ecological spaces and opening them up to communities for recreation (Snell and Fairbrother, 2010; Stevis et al, 2018).

Workers hold an important influence at key points and moments of the supply chains and flows that underpin the fossil fuel economy. This includes those working in shipping lanes and ports, oil refineries, power stations or construction sites. It is possible for these workers to exercise such influence to slow carbon-heavy industries to a halt: both to demonstrate their importance and as a tool of negotiation in disputes with employers. For example, in 2019, French energy workers staged power cuts that affected thousands of homes and businesses to protest government efforts to reform their pension scheme (Rose and Felix, 2019). The same is true in renewable energy, with workers retaining a significant role across the supply chains and flows that underpin an energy transition.

Policies of climate action and decarbonisation need to focus on creating secure jobs in a just energy transition, rather than just investing in renewable energy infrastructure itself. A key route to support local economies is to locate them as sites within the broader supply chains that support the decarbonisation found in sites of renewable energy generation. The route to secure, good, green jobs in many regions is found in the development of resilient and national supply chains. These green jobs are often stimulated and protected by government policy. In China's Yangtze River Delta, concerted central and regional government policy developed a domestic supply chain that supports the expansions of solar PV infrastructure: with industrial infrastructure, cheap electricity, and the presence of numerous suppliers (such as those in the glass industry) all allowing solar firms to buy products cheaply and quickly (IRENA, 2020b). These policies also ensure that the green jobs stimulated by a national renewable energy transition remain in China. Similar policies have been enacted elsewhere: governments in France, Taiwan and Turkey have all worked to ensure the use of domestic manufacturers is a pre-condition for offshore wind contracts (STUC and Transition Economics, 2021).

Whilst wind turbine blades are made domestically, the UK is heavily reliant on imported nacelles and turbines. The domestic supply chains that support national energy transitions remain precarious, with limited stipulations that prioritise their use by energy projects in the UK. There are some successes: for example, 87 per cent of the blades used in the Seagreen I project in the North Sea off Scotland are to be manufactured in the UK (Seagreen, 2021). However, many materials are manufactured elsewhere and imported. The absence of supportive supply chains can compound the problems facing green jobs, with the work associated with renewable energy projects becoming focused on construction or installation only. Once a project is complete, contracts expire, and jobs are lost as operations are restricted to maintenance and management.

The absence of concerted policies to support supply chains renders green jobs precarious and at risk in the UK. A case in point is Burntisland Fabrication (BiFab) in Fife and the Isle of Lewis, Scotland, which went into administration in December 2020. This followed the withdrawal of state support for the manufacturing of the jacket foundations needed for the Neart na Gaiothe offshore wind project and a failure to secure a contract to provide them for the Seagreen I project (the jackets for which will be built by Lamprell, based in the United Arab Emirates). In missing out on both contracts, BiFab lost future income. The Scottish government had already bailed out the company in 2017 to cover losses incurred from a contractor not paying BiFab for work completed. Due to this support, it argued that it was unable to help further in 2020, due to European Union regulations against state-supported industries.

The experience of workers at BiFab represents the necessity of supply chain protections in national green jobs agendas, with government-supported manufacturers in other countries able to undercut BiFab and other UK manufacturers. In 2017, under the threat of redundancies, BiFab workers staged a 'work-in' – continuing to work despite the company not having the funds to pay wages (Unite, 2017). Workers later occupied a plant at Methil, Fife to stop the closure of the plant and the loss of 1,400 jobs. Welders, scaffolders, crane operators, older workers, and new apprentices were present.

These workers highlighted how the loss of supply chains represents a series of broken promises to both them andt their communities. As Hazel Nolan, an organiser for the GMB Union, explained: 'They [the workers] are not just fighting for their jobs, their fighting for the whole community here and the Scottish economy' (Whitaker, 2017). The occupation did not save jobs, with the workforce reduced in numbers in 2017. While two of the three sites owned by BiFab were later bought by the Belfast shipyard Harland & Wolff in 2021 (creating close to 300 jobs), its demise exposed a reality of current green jobs agendas: that companies and jobs are often on a knife-edge because of a lack of state support (Hutcheon, 2021).

The BiFab occupation finds a parallel in action ten years earlier at the Vestas factory in Newport, Isle of Wight in 2009. In response to the planned closure of the plant, which manufactures wind turbine blades, and the loss of 525 local jobs, 25 workers occupied the administrative buildings on the Vestas site (Weaver and Morris, 2009). Management at Vestas argued that such a move, and the closure of another facility in Southampton (with the loss of 100 jobs), was due to a slowdown in demand. The resultant occupation lasted three weeks and inspired solidarity across the community (Perry, 2019). The action was, ultimately, successful. At the time of writing, the Vestas manufacturing plant remains on the Isle of Wight. The facility produced its 1000th offshore wind blade in 2021.

Events at the Vestas plant in Newport, the Isle of Wight, have an important symbolism in a just energy transition. While the action was rooted in challenging planned redundancies, it took on a wider meaning with production at Vestas having an important role in decarbonisation (Gibbs and Kerr, 2022). Central to the claims made by the occupiers was that the closure of the Vestas facility represented a departure from the pledges made by the government of Prime Minister Gordon Brown (2007–10) to address climate change (Treacy, 2010). Saving jobs became synonymous with saving the planet, with the 'green' jobs at Vestas deemed worth fighting for. As a group of Vestas workers argued in an online article:

> One of the questions that we all – campaigners, both environmental and trade union, and all working people – need to examine is whether we can let job creation, and the transition to renewable energy production that we need, rest on the short-term business decisions of private companies whose guiding principle is their bottom line. We argue that we cannot. We need to act as a public collectively, in our collective interest, including, if necessary, taking over plants and industries that cannot or will not deliver the change we need. (Ash et al, 2009)

While some environmentalist groups (such as Friends of the Earth and Greenpeace) were involved, they kept a relatively low profile in this disupute.

Other more radical environmentalist groups played a bigger role, staging solidarity gatherings, for example. However, the push to keep the plant open was led by the workers themselves. This worker action highlighted an important complexity of energy transition in the UK: namely that climate action and the workers' futures were intricately entwined. Far from being focused on wages and employment, workers at Vestas were aware of the role their jobs have in broader climate action – and sought to protect such work and its significance (Hampton, 2015).

Workers can take a key role in defining a just energy transition. An important example of workers pushing for better, greener work can be found in the 1976 Lucas Plan. This document originates from the work of a group of workers at 15 plants owned by Lucas Aerospace Corporation in the UK that were threatened with mass redundancies. At Lucas, the workforce had been reduced from 18,000 to 13,000 between 1970 and 1974, likely precipitating further job losses (Beynon and Wainwright, 1979). It is from this uncertainty that the Lucas Plan emerged. Shop stewards consulted with members and constructed a plan around the skills and aspirations of the existing workforce. The final plan included designs for 150 products, including hybrid-electric vehicles and heat pumps (Lucas Aerospace Combine Shop Steward Committee, 1976). Across the Lucas Plan, workers put forward an alternative vision to save jobs by reorienting production around goods that would help people, communities, and the planet (Smith, 2014).[11] Rather than manufacturing military hardware, they sought to make dialysis machines and wind turbines. While the Lucas Plan was rejected by the employers, it provides an important example of a group of workers, who held significant industrial expertise and experience, coming together and not only seeking to secure long-term employment but ensuring that such work was useful and beneficial to their community. In a context of climate change and growing inequality, the Lucas Plan presents a beacon in a just energy transition: of workers seeking to take control of manufacturing plants and processes and dedicating their labour to social and environmental good. Rather than a business model based on shareholders and profits, the Lucas Plan called for a worker owned industry to produce goods that helped people.

While there remain some trade unions that oppose climate action, many show direct support for decarbonisation policies (Thomas and Doerflinger, 2020).[12] Union members are, on average, more likely to display pro-environmental attitudes than other members of the population and labour unions are increasingly active at global climate summits (Vachon and Brecher, 2016; Allan, 2020). Workers' groups have provided important support for and built coalitions with environmentalists when action also ensures that jobs are protected and working conditions improved. This is not necessarily a new process: the International Union, United Automobile, Aerospace and Agricultural Implement Workers (UAW) contributed financially and logistically to the first Earth Day in 1970 (Stevis, 2019).

These historical examples of worker action are important – however, it is necessary to not only rely upon past examples when defining what a just energy transition might be. In today's energy policies, trade unions, representing workers in many different sectors, have put forward their own visions of decarbonisation to ensure that 'green' jobs are 'good' jobs. In the United States, Climate Jobs Illinois (part of a nationwide effort led by the Climate Jobs National Resource Center) and the Illinois Clean Jobs Coalition have pushed for one of the most pro-worker energy bills enacted in the US: The Climate and Equitable Jobs Act, approved by the Illinois General Assembly in 2021. This legislation ruled that, to get state support, renewable energy projects must work with trade unions, pay good wages, and demonstrate action to widen the diversity and inclusion of their employee pools (Illinois Government, 2021). In addition, a Clean Jobs Workforce Network Hubs programme was established, creating 13 hubs across the state with funding of $80 million per year to support training for low-income communities. This plan was formed through close work between legislators, community groups, trade unions and other groups – highlighting the significant potential of such collaboration and dialogue.

In the United States, workers have taken a stand to both create and protect skilled, well-paid jobs that are the cornerstone of decarbonisation. In New York, trade unions have set up the Climate Jobs in New York organisation, representing over 2.5 million workers in advancing an explicitly pro-worker and pro-climate agenda. A key element of its action is to focus on securing what is known as a project labour agreement in new wind projects in the state. These are highly important mechanisms and represent a company agreeing with local unions to employ union workers and pay union wages throughout a project's construction period.[13] It may also incorporate policies related to apprenticeship training programmes to ensure both a constant pipeline of labour (good for the company) but to provide new opportunities for younger or under-employed workers (good for the worker). In California, project labour agreements have enabled the use of a state-wide apprenticeship system in renewable energy projects (Luke et al, 2017). This has also increased the accessibility of the jobs created: at solar projects under construction in Kern County, California, 43 per cent of entry-level workers came from communities that were on lower incomes and/or experienced higher-than-average unemployment (Luke et al, 2017).

At the centre of project labour agreements in the United States is the provision of a dialogue, in which the company maps out the project lifespan and communicates with the union decides how many workers they will employ at each stage and in each year. In Maine and Connecticut, legislation has stipulated that labour standards must be in place on all renewable energy projects above a certain megawatt. In Connecticut, policies have been passed to ensure that renewable energy projects provide a fair wage to workers and establish workforce development schemes (Connecticut House Democrats, 2021). This lobbying stems from a realisation that trade unions were often excluded from previous energy policies, resulting in renewable energy projects importing cheap, out-of-state labour (Roberts, 2021). These examples highlight how trade unions have their own vision of decarbonisation and one that does not necessarily require environmentalist input. Climate Jobs New York defined their own pro-worker approach without speaking to other groups – not because their input was unwanted but due to the central importance of placing the workers' voice at the core of any transition[14]. The unions were clear on their own position and on the importance of worker protections and project labour agreements, to ensure that all jobs created were unionised, well-paid, and secure.

Rule 5: Ensure the participation and inclusion of workers

A just energy transition requires new policies that prioritise the benefits of green jobs associated with renewable energy, rather than purely focusing on the products they create. 'Green' jobs can often be precarious or poorly paid, with limited routes to move to them from other work leading to the exclusion of many. Pledges of decarbonization exist in the short term but the solution to how these policies change work, livelihoods, and employment is not. It involves long-term thinking about changing economies, skills development and training programmes, the development of broader employment in supply chains, and the empowerment of workers. Positive benefits of companies training and employing workers in well-paid, secure jobs with good working conditions include the spending of wages in local economies, the support of livelihoods for workers facing insecurity in carbon-heavy employment, and the provision of new training and skills.

Policies that support green jobs can involve more than just traditional top-down instruments of industrial policy (such as subsidies or regulation) but can, instead, include more-collaborative approaches between public institutions and services, small and medium sized businesses, and local communities (Rodrik, 2022). Renewable energy schemes aid the processes of community wealth building discussed in Chapter 5 through their roles as job creators. New jobs and sectors can be stimulated, with projects ensuring that work (and the wages paid) are secured by local contractors and workers. Effective policies in this vein might include the specific needs of small and medium-size businesses in a local area and provide training to workers to fulfil these needs (Bartik, 2020; Rodrik, 2022). Local and regional authorities in the UK can influence adult education and skills programmes in this way. As a result, they can play an important part in uncovering, defining, and shaping the skills base required for a just energy transition. A regional approach to green skills might use adult education funding and services to provide workers with training in installing heat pump technologies and provide direct support to schools and colleges to not only train the workers required for expansive retrofitting– but also coordinate efforts to ensure that this process drives greater diversity in the sector.

Work must also be done to increase the attractiveness of new green jobs for both those workers who are new to the job market and those who are either looking to move from or have lost jobs in carbon-heavy industries. To do so, trade unions and worker organisations must have a part to play in co-developing plans and regulations to ensure that new, green jobs are good jobs (Emden and Rankin, 2022). Worker action at Vestas and BiFab demonstrates how workers in green jobs can be aware of the role of their labour in wider climate action. There are other examples of this. The construction of the wind farm at Blyth, described at the beginning of this chapter, has roots in the Northumberland Energy Workshop, a small worker's cooperative that provided a space for engineers and workers in local industries to support wind energy projects (Dawley, 2014). A similar process to the Lucas Plan occurred in 2021 at a GKN plant in Birmingham, where workers went on strike in response to plans to close the plant. This strike was in response to the company's refusal to hear an alternative workers-led plan to pivot manufacturing at the plant to electric vehicles (Diski, 2021b). Both proved unsuccessful. The Northumberland Energy Workshop ultimately folded due to a lack of state support and the GKN plan was rejected by employers.

Despite these frustrations, episodes like this provide important examples of what a just energy transition might look like: with workers and communities protected, influencing the direction of change, and putting their skills towards work that will benefit many. Such moves should be facilitated further, with a just energy transition placed at the centre of discussions around expanding renewable energy supply chains. What is required is a commitment to broader action with a wider, longer-term vision of skills and apprenticeships providing pipelines to work that is better than what was available before – both for those leaving fossil fuel workers and younger people now entering employment. Green jobs provide a valuable opportunity for the building of a coalition between those pursuing decarbonisation and those living at its sharp end. Workers should be retrained and protected and given influence over the work that they do.

The global supply chains that support renewable transitions don't just affect communities in Blyth and California. They extend across countries and communities, affecting people and groups across different sites and places. As a result, they become inscribed with the global injustices of extraction, pollution, and exclusion that necessitate a focus on and approach of climate justice.

8

Global

The small village of St Dennis can be found a short journey inland from St Austell on the south coast of Cornwall. Named *Tredhinas* in Cornish and once home to an iron age fort, the village lies at the centre of the county. It is equidistant from the River Tamar to the east and Land's End to the west, and equally between the English Channel to the south and the Atlantic Ocean further north. Perched up on a hill above the village is St Denys Church. From this viewpoint, you can witness a landscape marked by centuries of toil and excavation. The village is nestled in the 'Cornish Alps', a collection of peaks formed by waste from clay-mining in the region. The landscape is rugged, uneven, manufactured and mined. Some peaks are table-topped, others rough and pointed. In the distance are vast cuts to the earth that have created new white cliffs that overshadow the landscape. These marks stem from the region's extractive history. In the early 19th century, this area had the biggest known deposits of kaolin in the world. This material, formed of decomposed granite, is used to make white porcelain, paints, and dyes among many other things. Its mining came to be a key part of the regional economy, employing thousands of residents to remove the clay from the ground, process it, and transport it across the globe. With every tonne of 'Cornish Clay' mined, a vast amount of waste emerged. Thrown into heaps, it was piled up into the mountains that became the 'Alps'. Clay mining also dug deep, leaving holes in the landscape, some of which now hold blue pools of toxic water.

While the extraction of kaolin from this landscape still occurs today in the nearby town of Par, the Cornish Alps now host new prospectors who seek a material central to any energy transition. Lithium is used to manufacture solar panels, and the batteries used by both electric vehicles and to store the energy created by renewable sources. Two companies, Cornish Lithium and British Lithium, are leading this charge into the ground. Newly drilled boreholes are exploring what is possible. The history of mining in the area retains an importance: the identification of pilot sites in this new venture has stemmed from the close analysis of historical, hand-drawn maps that showed who had dug where and how deep to create a roadmap of where to dig and how to avoid crashing into tunnels that had been built generations before (Bernal, 2021). The discovery of lithium in this region and its removal from the ground could well be transformative: it has been estimated that enough lithium might be mined to manufacture more than 500,000 electric car batteries a year, employ 8,000 locals, and pay £1 billion into the local economy (Bernal, 2021).

Across the globe, communities like those in the Cornish Alps, are finding themselves becoming key sites in the supply chains that support energy transitions happening elsewhere. It has been estimated that the expansion of renewable energy required by 2015 Paris Agreement, which aims to limit the global temperature rise to 1.5 degrees Celsius, might require as much as 34 metric tonnes of copper, 50 million tonnes of zinc, 162 million tonnes of aluminium, and 40 million tonnes of lead (Hickel, 2019). All of this must come from somewhere. It must be detected, extracted, shipped, processed, and manufactured into renewable energy technologies. It is through this demand for minerals that renewable energy technologies leave footprints, linked to their deep roots in global supply chains that include mines and glassworks, smelters, and semi-conductor manufacturers (Mulvaney, 2013; 2019).

A renewable energy transition and, with it, energy justice occurs across a long chain of sites and communities from the place of energy generation to the point of use and beyond (Hernández, 2015). These technologies are the product of complex global supply chains that extend across borders and between communities and their lifecycles raise questions for a just energy transition – from their removal of key materials from the ground to the afterlives of the technology themselves when they are replaced.

In the context of global climate action, the materials that underpin renewable energy technologies hold a geopolitical significance. The geographies of where these rare earth metals and minerals are located can make them precarious or place them at risk. The Russian invasion of Ukraine in February 2022 coincided with skyrocketing prices of nickel, due to Russia holding a vast proportion of global reserves of a key resource in the manufacturing of electric vehicles (Pickrell, 2022). National governments have also sought to develop greater control over the mining and trade of these minerals to ensure that the benefits are invested within the country itself. In 2022, Mexico's House of Deputies voted to nationalise its lithium resources (Agren and Stott, 2022). The same year, Indonesian government officials floated the idea of creating a new international mechanism, similar to the Organization of the Petroleum Exporting Countries (OPEC), to secure control over its nickel, cobalt and manganese reserves (Dempsey and Ruehl, 2022). Other national governments have sought to find new domestic sources of these materials and to secure their supply. Portugal has started exploring for lithium in several areas, hoping for a site to supply lithium across the European Union (Kijewski, 2022). In the United States, Nevada has emerged as a key site of exploration and potential extraction (Flin, 2021). For the UK government, it is the lithium found in Cornwall that could gift new levels of resource independence in decarbonisation.

This chapter explores how a just energy transition must include the spaces and communities that host, produce, and provide renewable energy technologies for decarbonisation elsewhere. A first step in doing so is understanding both energy transitions and the harms that they might create as 'transboundary', flowing across borders and connecting communities across supply chains (Healy et al, 2019). This requires understanding the full lifecycle of renewable energy infrastructure and its afterlife as including several points, namely of extraction (where and what the technology comes from), production (where it might be sited), and waste (where it goes once it is replaced).

Embodied injustices and geographies of responsibility

Fossil fuels have always carried an embodied injustice, inherent in the material from the moment it is pulled from the ground (Healy et al, 2019). These include the health impacts suffered by those who mined coal, such as pneumoconiosis ('black lung') caused by the inhalation of coal dust. Or the lost livelihoods of those impacted by oil spills. Or the state-led violence that underpinned the expansion of fossil fuel extraction. Renewable energies also hold significant injustices across the supply chains and production lines that manufacture, host, and later dispose of them. This is heavily linked to the minerals and materials required in the manufacturing of renewable energy technologies, which are formed of numerous components that are brought together and combined. A wind turbine requires zinc, manganese, and copper (IEA, 2021a; Glüsing et al, 2021). This technology, a key part of any energy transition, may, at times, require nine times the mineral resources to be manufactured than a gas fired power plant (IEA, 2021a).

The demand for minerals in energy transitions is also driven by the need for energy storage. While hydrocarbons, like oil and coal, can be extracted and held in storage until they need to be burnt for energy, renewable energies are more variable. They generate electricity intermittently, and, as a result, an energy transition requires investment in grid-scale storage, to ensure electricity is reserved for moments of high demand. Both reservoir and pumped-storage hydropower are technologies widely used for moments of high demand (releasing vast amounts of water through its turbines when needed). However, the most-scalable technology for this purpose comes in the form of batteries. Grid-scale storage is a technology that would charge large batteries from the grid when needed, store electricity and then release it when supply might be outmatched by demand, such as in the evenings or during heatwaves. Recent years have seen advances in this technology that would enable the development of large battery sites. Battery storage facilities in California, a world leader in the technology, might be able to store enough electricity that, when released, could power around 300,000 homes for four hours (Katz, 2020). All of this would require a vast amount of lithium to manufacture the battery technologies used.

While we often think of renewable energy technologies as purely existing at the sites at which they generate electricity, their demand for minerals and metals highlights a need to include stages of manufacturing, construction, consultation, maintenance, and disposal in our understanding of energy transitions (Heffron and McCauley, 2014; Mejía-Montero et al, 2021). It is at these sites that an important contradiction of renewable energy transitions is illuminated: namely, that the manufacturing of technologies deemed 'green' has significant social and environmental impacts (Bainton et al, 2021). These impacts are often overlooked in energy policies, due to them occurring a world away from where the technology is used and the policy makers who might see energy transitions as the reduction of greenhouse gas emissions only. However, they represent an important process of cost-shifting similar to the energy peripheries discussed in Chapter 3: with the negative consequences of renewable energy felt in one location, but the benefits enjoyed elsewhere.

There is a need for a just energy transition to understand and address the experiences of those who work to extract, manufacture, or recycle renewable energy technologies and those communities who live with the impacts. To do so requires an approach of cosmopolitan justice, which highlights a universal approach to justice by emphasising that all individuals and countries share are bound by a shared and equal moral worth and responsibility to others. Efforts to ensure a global sense of justice might include policies focused on the alleviation of poverty, the provision of basic freedoms and/or human rights, the emergence of new social structures, or global equality between countries themselves (Sen, 1999; Brock, 2009; Nussbaum, 2011). Climate change itself is a problem of cosmopolitan and global justice. Local emissions in one part of the world have important global consequences and impact other communities that have far less responsibility for them. Coal is mined and burned in certain countries, emitting greenhouse gases, warming the planet, and leading to sea level rise that displaces populations elsewhere. A just climate change response, in cosmopolitan terms, might have three pillars: a lowering of global emissions, a prioritisation of adaptation and resilience to impacts, and the rapid systemic change required to move away from the fossil fuel regime that has caused the problem in the first place (Dietzel, 2022). For this to happen, there must also be a reckoning with both the historical responsibility for emissions (including the United States, UK, and other countries in Europe), as well as the links between climate justice and historic and entrenched injustices associated with imperial expansion and colonial subjugation. These processes, if done right, can be transformative. An approach of cosmopolitan justice requires a shift in how we see the world by highlighting how we have responsibilities linked to how our actions and emissions have impacts across borders (Delanty, 2014).

In this sense, a just energy transition requires a rethinking of how local and national decarbonisation plans exist within a global supply chain of goods, labour, and materials. Our energy demands (and how they can change) have global consequences. The boundaries of our communities, behaviours, and decarbonisation are broader and more porous than we might think at first. This adds a new geographical dimension to a just energy transition, with distributive and procedural injustices reaching beyond our immediate communities. Local and national energy transitions are tied to wider 'geographies of responsibility' that link together communities across borders, the energy grid, and supply chains (Walker, 2009). This is evident in the materials and labour needed to manufacture the renewable energy technologies that are driving decarbonisation in some places yet are also driving important changes and impacts in others.

Manufacturing renewable energy technologies

A key mineral in a renewable energy transition is lithium. This silvery-white soft metal is used to manufacture the batteries that will be used to store electricity in smaller domestic batteries in homes and larger storage facilities across electricity grids. It is also layered with silicon semiconductors and glass to make solar photovoltaic panels that are placed on rooftops and in fields. Lithium exists in numerous forms and can be mined in different ways. In some places, it is found embedded within other minerals in rock or clay, which are removed from the ground and processed. In other locations, it exists in the brine on salt flats, where it is extracted by dissolving the saltwater to leave the mineral behind.

Global demand for lithium exposes how the current model of decarbonisation is recreating unequal dynamics across different parts of the world, through the shifting of the costs of a renewable energy transition from one place to another. Our need for lithium is expanding: to meet the Paris agreement goals, demand will likely increase by close to 90 per cent (Singh, 2021). The reserves needed to fulfil such requirements are concentrated in only a few countries. Three countries in South America hold half of the global reserves of lithium between them: these are Bolivia (an identified 21 million tonnes), Argentina (19 million), and Chile (9.8 million) (USGS, 2022). Much of the global attention on lithium extraction and trade has focused on these countries and the 'lithium triangle' region that cuts across their borders, encompassing the Atacama (Chile), Hombre Muerto (Argentina), and Uyuni (Bolivia) salt flats. While at varying stages, the lithium found in these landscapes has provided an important driver of economic policy with political narratives around lithium extraction promising prosperity for nearby communities and new jobs to support global energy transitions (Voskoboynik and Andreucci, 2021).

To extract lithium from the salt flats, mining companies suck up the brine beneath the ground, sometimes at a rate of 1,700 litres every second (Riofrancos, 2019). The brine is then evaporated through a patchwork of successive yellow-green pools, leaving behind a lithium-rich salt. This can take months to evaporate, with the mixture of lithium and other minerals (such as manganese and potassium) then transferred to other evaporation pools to further filter out the lithium carbonate needed. This process takes a long time, often over a year, and the installations that complete these processes are often vast. In Chile's Salar de Atacama in 2017, these sites were estimated to occupy a total of up to 80 km^2 of land – expanding in size by over 7 per cent per year (Liu et al, 2019).

The process of evaporation deployed in these ponds is financially cheap but it demands vast amounts of water. It has been estimated to take up to 225,000 litres of water to 'mine' a tonne of lithium this way (Katwala, 2018). This has important consequences for many living in the region, because of its impacts on the availability and quality of water (Giglio, 2021). Access to water is a necessity for local communities and their livelihoods, as well as for local biodiversity but drilling underground has risked the mixing of the salty brine with the water that is used by those who live there, contaminating it to such an extent that it can no longer be used. Communities in Argentina have complained that lithium extraction has contaminated streams used to serve livestock farming and the growing of crops (Garcés and Alvarez, 2020). In other areas, lithium mining has coincided with a lowering of the water table, worsening the changing flows of rivers caused by climate breakdown (Riofrancos, 2019). These impacts on the region's water are not just hydrological issues. Many Indigenous communities in the region have deep historical, cultural, and cosmological relationships with the region's water systems, which have been disrupted and dislocated by lithium mining (Jerez, 2021).

The mining of lithium has transformed the region. Water reserves have been contaminated, as have the soil and air (Bolger et al, 2021). Lithium mining companies are permitted to extract defined amounts of water per day by the government and, while they have pledged to compensate local communities, this is often inadequate to address the irreparable damage to the landscape, community, and the links between the two (Garcés and Alvarez, 2020).

In the lithium triangle, one water use has been prioritised over others, with the demands of lithium extraction sites put before the needs of local communities (Giglio, 2021). Elena Rivera Cardoso, the President of the Indigenous Colla community in the Copiapó commune, northern Chile has described how:

> We used to have a river before that now doesn't exist. There isn't a drop of water ... And not only here in Copiapó but in all of Chile, there are rivers and lakes that have disappeared – all because a company has a lot more right to water than we do as human beings or citizens of Chile. (Greenfield, 2022)

Patterns of land ownership have also been redefined in these processes, leading to tensions between Indigenous land rights and the spaces where companies want to extract lithium (Sanchez-Lopez, 2021). In many projects, particularly those in Chile, communities were not consulted about impacts, nor do they necessarily participate in the decision-making that leads to the projects themselves. Many members of the community have struggled to adapt to the changing landscape and region. It is happening too fast and too dramatically, leading to more and more people moving away to find new lives and livelihoods beyond the lithium triangle.[1]

The salt flats of the lithium triangle are key strategic sites in a global energy transition. As a result, they have been transformed – as have the lives of those who live there. Lithium mining in this region has fractured communities, disconnecting them from both one another and the landscape around them, and altering patterns of livelihoods and everyday life (Jerez et al, 2021). Yet the communities of the lithium triangle are excluded from the benefits. In Chile, lithium extraction in the Salar de Atacama has been linked to an influx of workers into the region but has not led to new jobs for residents, with the proportion of local labour involved in mining in the region dropping from 52 per cent to 18 per cent between 2002 and 2017 (Liu and Agusdinata, 2020). Despite initial claims from the government of President Evo Morales (2006-2019) that Bolivian lithium would be processed and managed through a state-run industry and domestic supply chains building electric vehicles, lithium mined in Bolivia is now mostly exported as a raw material to the global market (Achtenberg, 2010; Kingsbury, 2021). It is shipped in the form of a grainy powder to be processed in other countries and by workers elsewhere.

This represents an episode of cost-shifting, in which the communities of the lithium triangle bear the impacts linked to the extraction of a material that is shipped, manufactured, and used by others elsewhere. This cost-shifting can also play out in geopolitical tensions. In the lithium triangle, both US and Chinese companies have competed to secure the resource and exclude their rival. In Bolivia, political upheaval and the removal of Evo Morales from the Presidency in 2019 were characterised by Morales and his supporters as being part of a foreign-backed coup to secure the country's lithium reserves (Anderson, 2020; Morales, 2020). These tensions are rooted in a particular set of political and economic conditions that have historically placed these regions and communities at the periphery of global trade, with many highlighting the significant parallels between the extraction of trade of lithium in South America and the pillage of the same region both by Spanish colonists centuries ago and private companies in the 20th century (Farthing and Fabricant, 2018; Jerez et al, 2021).[2]

This shifting of costs and benefits is evident elsewhere in the current model of energy transitions. Many of the companies mining minerals that are key to energy transitions, such as lithium, copper, manganese, and zinc, have been accused of human rights abuses (BHRC, 2019). An important example can be found in cobalt, which is used to manufacture the batteries that store solar and wind generated electricity, as well as those used in electric vehicles and smartphones. Most of the global reserves of this mineral are found in Zambia and the Democratic Republic of Congo (USGS, 2022). It is at the sites of extraction in these countries that the mining of cobalt has important impacts. In the Democratic Republic of Congo, it has been linked to child labour and abusive working conditions (Sovacool, 2021; Bernards, 2022). Exposure of workers to toxic pollution in these mines has also been understood to have led to birth defects in infants born in nearby communities (Amnesty International, 2020). With workers toiling with limited protective equipment, their exposure to pollution is heightened. In 2019, a class-action lawsuit was filed in the United States against several large companies by parents from the Democratic Republic of Congo who sought to hold these companies accountable for child labour practices at cobalt mines in the country (O'Donoghue, 2021). Plaintiffs argued that their children, who worked illegally at mines owned by the Glencore and Zhejiang Huayou mining companies, were sometimes paid as little as £1.50 a day (Kelly, 2019). Some were killed in workplace accidents, others suffered from life-changing injuries. The metals extracted at these miners were alleged to have been sold to companies in the United States, including Apple, Google, Dell Technologies, Microsoft, and Tesla, all of whom were named in the lawsuit filed on behalf of 14 parents and children working in cobalt mines (Kelly, 2019). In response, the companies argued that they had no control over their global supply chains. This is a similar argument made by other large multinational companies facing criticism for human rights violations at the local sites that manufacture goods that are shipped and sold on the global market. In doing so, these companies obscure the important webs of injustice and geographies of responsibility present within global supply chains and the extraction of minerals to serve new technologies used elsewhere.

The embodied injustices of renewable energy technologies can also be found at the sites where they are made and manufactured. Solar panels, in particular, have been linked to human rights violations and exploitative labour practices. An example can be found in polysilicon, high-purity grade silicon which is sliced into a thin film and then layered to make solar panels. Around 45 per cent of the global supply of the polysilicon used in solar panels is understood to be produced in the Uyghur region of the People's Republic of China (Murphy and Elimä, 2021). In this region, the government of China has forced Uyghur and Kazakh citizens into what is labelled 'surplus labour' programmes. An estimated 2.6 million people have been enrolled in such schemes (Murphy and Elimä, 2021). While the Chinese government claims that these schemes are voluntary, there is growing evidence that those enrolled are unable to leave these schemes (Amnesty International, 2020). In 2021, the European Center for Constitutional Rights (ECCHR) submitted a criminal complaint against several global clothing companies, such as Nike and Patagonia, accusing them of being directly or indirectly complicit in the use of forced labour in this region (ECCHR, 2021). 11 companies manufacturing polysilicon in the Uyghur region are reported to have participated in forced labour transfers and, as a result, provide key links between global solar energy technologies and the use of coerced workers in supply chains (Murphy and Elimä, 2021).

The experiences of those workers tasked with mining cobalt and manufacturing solar panels In the Democratic Republic of Congo and Xinjiang represents a global process of cost-shifting in decarbonisation. Cheap or forced labour is used to drive down the costs of manufacturing renewable energy technologies but has forced the negative impacts on health, wellbeing, and freedom onto communities living a great distance from those who enjoy the benefits of an energy transition. These places, communities, and workers are central cogs in global decarbonisation. They toil at key sites that create the technologies that global climate action relies on – yet they gain little from an energy transition. Instead, their landscapes and livelihoods are utterly transformed to provide the benefits of solar power to those living elsewhere. The experience across these supply chains suggests that solar panels are like any other global commodity, linked to extensive social and environmental impacts in regions that become the sacrifice zones for global supply chains (Brock et al, 2021).

Global energy flows

The geopolitics of energy transitions, evident in the global trade of lithium, cobalt and nickel, is also seen in how the potential to generate electricity from renewable sources is not spread equally across the globe. Countries with a high potential for solar energy include those in North and Sub-Saharan Africa, as well as Afghanistan, Argentina, Iran, Mexico, Mongolia, Pakistan, and islands in the Pacific and Atlantic Oceans (ESMAP, 2020).

These places have low seasonal variations in the amount of sunshine that they get, meaning that they can generate solar energy all year around. Significantly these countries are home to about 20 per cent of the total global population but hold some of the lowest levels of electricity access globally (ESMAP, 2020). Investment in solar energy infrastructure in these areas can increase energy access and allow many households and communities to gain energy security and independence. This can be through rural electrification programmes, in which smaller projects are built locally to provide nearby homes, businesses, and communities with new forms of secure, low-carbon electricity. In many countries, like India, these schemes have converged with broader social priorities, such as those protecting livelihoods or expanding irrigation networks for farmers (Palit and Kumar, 2022).

While renewables have allowed for the expansion of energy access in many countries, larger-scale facilities have become linked to broader global flows of energy and, with it, political power, and injustice. There is a global imbalance in the potential to use solar power to provide energy to national populations. In Ethiopia, the covering of a mere 0.003 per cent of the country's total land area with solar panels would generate enough electricity to meet its population's current energy demands. In France, the equivalent area would be closer to 1 per cent (more than 5,000 km^2) (ESMAP, 2020).[3]

The global geography of solar power potential has led to renewable energy projects becoming tools of diplomacy and global politics. This is evident in the case of Morocco, which has expansive solar energy potential and limited seasonal variation. The Moroccan government, led by King Mohammed VI (1999 onwards) has staged a series of ambitious plans leading to the rapid expansion of renewable energy capacity through investments in both wind and solar. This was, in part, driven by a desire to wean the country from fossil fuel imports and allow for future energy security. Morocco is now home to some of the world's largest concentrated solar plants, such as the Noor-Ouarzazate complex which extends out by close to 25 km^2.

Renewable energy projects in Morocco have been celebrated as ensuring national energy independence. However, within Morocco, these facilities represent the appropriation of land and exclusion of local communities (Rignall, 2016). Many renewable energy projects in the country are clustered in the region of occupied Western Sahara. This region was a Spanish colony but was annexed by Morocco between 1975 and 1979. Claims that Western Sahara historically belonged to Morocco and neighbouring Mauritania have been disputed by Saharawis – who set up a government-in-exile that retains control over one-fifth of the area. Those Saharawis who did not flee in the 1970s have been cut off from this territory by a large sand wall. A long civil war ended in 1991 with a UN-brokered ceasefire that included an referendum. This vote on the region's autonomy and future is yet to take place.

Morocco's positioning as a global leader in renewable energy is linked to this occupation of Western Sahara, displacing residents, and communities, and described as building projects with impunity, without consultation or consent from the Saharawi population.[4] This region contributes 18 per cent and 15 per cent of Morocco's national wind and solar installed capacity respectively (Allan, 2021). This is expected to increase, likely leading to further authoritarianism and occupation greenwashed by renewable energy (Allan et al, 2021; Western Sahara Resource Watch, 2021). The international perception of renewable energy as 'green' limits the space for local opposition against the impacts of these projects (Ryser, 2019; Allan et al, 2022).[5]

In October 2022, the European Union and Morocco signed a 'green partnership', further linking energy transitions in the former with renewable energy generated in the latter. With Morocco's coastline 15 km from the Spanish mainland, such cooperation may lead to the electricity generated in Western Sahara being transmitted to the European continent. This is not a new idea. The expansion of Moroccan renewables have led to many international observers raising the possibilities of transmission lines carrying electricity from Moroccan arrays to countries north of the Mediterranean Sea. The DESERTEC project, discussed for many years, aimed to export electricity from renewable energy plants in the deserts of North Africa to Europe. Energy connections to Morocco and neighbouring Tunisia remain seen as a key route to the European Union meeting its targets for emissions reduction (Bennis, 2021).

Generating electricity in one place and using it elsewhere is not a new occurrence. Transmission power cables already run great distances. The North Sea Link runs over 700 km from the Kvilldal hydropower plant in Southern Norway to the UK. However, the scale if these connections is changing. In addition to DESERTEC, the Xlinks Morocco-UK project has emerged. This scheme seeks to build four 3,800-kilometre-long cables between Morocco and the quiet village of Alverdiscott in the UK (Xlinks, 2022). In doing so, it will transmit 10.5 GW of electricity generated by solar and wind facilities in the region of Guelmim Oued Noun, Morocco (the southeastern part which is in Western Sahara) to the UK. In August 2022, planning permission was granted in the UK for the factory that would be tasked with building these transmission cables (Morby, 2022).

Through these expansive connections, renewable energy use in European countries may well become implicated in the broader patterns of land grabbing and human rights abuses, validated by claims of clean and green energy (Allan et al, 2021; Dunlap and Laratte, 2022). In Western Sahara, access to electricity is often defined by the Moroccan regime, with reports of districts with larger Saharawi populations being subject to more blackouts than others (Allan et al, 2021). Energy affordability can also be a problem, with some spending up to 60 per cent of their monthly income paying for electricity (Allan et al, 2021). Those living in Western Sahara have witnessed their wealth of solar power being used for the benefit of those living elsewhere. Solar power has not led to increased energy access or affordability, it has become linked to broader global flows of geopolitical influence. While the 'green' energy of Morocco might be celebrated, its export will represent another significant example of the cost-shifting of global energy transitions across international borders and how some communities bear the costs of infrastructure that provide benefits (of energy supply, security) to communities elsewhere while being excluded from any such gains.

Disposal and waste

Renewable energy technologies appear relatively modern and, as a result, might be expected to last a long time into the future. However, many technologies that underpin today's energy policies can soon become redundant. A smart meter is a technology used in households that measures energy usage and transmits this information back to the energy supplier in real time, helping to reduce demand. As of March 2022, 28.8 million homes and small businesses in the UK had smart meters, around 51 per cent of the total (BEIS, 2022d). This technology is formed of a microcontroller unit (a small computer on a single computer chip), a power system, and a screen that displays energy use to the user. As new and better technologies emerge, the old meters will need to be disposed of. Many first-generation smart meters are dependent on 2G and 3G signals, a technology that is already becoming redundant and will be switched off by providers in the near future (Woods, 2021). As this happens, many devices will become obsolete and need to be replaced by 2033 (Walne, 2022).

These older smart meters will likely soon be added to the piles of electronic waste (or e-waste) that workers sort through and recycle. The UK is one of the biggest producers of e-waste in Europe, with each person producing an average of 23.9 kg of e-waste in 2019 (Malloy, 2021). However, a large amount of this is not processed and recycled within the UK itself. Instead, it is often stuffed into shipping containers and sent elsewhere. It lands in countries like Ghana, where people and communities process it and are exposed to the toxicities that such a process creates. At places like Agbogbloshie in Accra, Ghana, workers toil to process phones and computers by hand, drawing out copper wiring and other precious metals to sell to scrap metal dealers (Kwan, 2020). This work exposes them to toxicities caused by the mercury, lead, cadmium, or other materials that might be found within an item's casing. The exposure to such harm at Agbogbloshie is discriminatory: migrant workers are often given the worst jobs due to their ethnicity or religion (Sovacool et al, 2020).

The risks faced by those processing e-waste in places like Agbogbloshie provides a cautionary tale for renewable energy technologies and the potential cost-shifting of waste from one place to another. The first generation of solar panels, manufactured in the 1980s and 1990s, are reaching the end of their lifespans. It has been estimated that the waste from solar PV may reach as high as 78 million metric tonnes by 2050 (IRENA, 2016). By 2030, the US alone may be decommissioning and replacing 1 million tonnes of solar panels per year (NREL, 2021). All of these must go somewhere – and they must be disposed of safely.

The final stages of renewable energy technology's lifecycles, if done poorly, can lead to health impacts for those tasked with processing them. The lithium and silicon in solar panels are hazardous to people through exposure and the environment through polluting soil or water. Women exposed to lithium may pass the toxins onto nursing infants through their breast milk, breathing in nickel can cause asthma, and cadmium, found in the solar photovoltaic cells, can be carcinogenic when in dust form (Aman et al, 2015; Kumar and Turner, 2020).

The need to recycle renewable energy technologies is not new nor unexpected. It has been discussed for decades (Fthenakis, 2000). However, there remains a need to expand safe recycling facilities of these technologies and to hold a for a broader discussion about producer responsibility in the future. This would involve more efficient and sustainable management of the metals, minerals, and resources used in the manufacturing process and the broader life-cycle of the technology itself (Xu et al, 2018). It is also in terms of creating new eco-friendly designs and ensuring the use of recycled materials through circular economy approaches (Mirletz et al, 2022). These approaches would lead to renewable energy technologies like solar panels being reused, refurbished, or recycled to ensure that they (and the materials that they contain) can be used for as long as possible. This might include approaches that seek to reuse solar panels in new places and on new rooftops when they are deemed to be too old to be used elsewhere. Or it might involve policies that regulate manufacturing to ensure that they can be taken apart, recycled and remanufactured. For example, the European Union's 2019 Waste of Electrical and Electronic Equipment Directive has dictated that all producers of solar PV panels in the region must also finance the costs of collection and recycling panels at the end of their lifespans. This represents a new form of responsibility, with those manufacturing panels being liable for this technology's impacts, even after their immediate role in making and selling them is complete. Recycling these panels, rather than letting them go to waste, would allow for the retrieval of a huge number of raw materials that can be reused by other facilities and new technologies. The potential value of these recycled materials could reach as high as US$15 billion (£12.2 billion) and could be used to produce 2 billion new panels in 2050 (IRENA, 2016). This is a vast amount and would stimulate an expansive secondary market.

The first wind power projects are also starting to reach the end of their planned lifetimes. Wind farms are generally provided planning approval with a time limit. As this limit comes up, the options are for the projects to either be 're-powered' (continuing to be used) or decommissioned.[6] In the UK in 2021, 112 wind farms were aged over 15 years old (17 per cent of the total number of wind farms in the country). As a result, the infrastructure (and the communities living nearby) will face discussions of 'what next?' in the relatively near future. With an estimated 80 per cent of the environmental impact of wind turbines found in their manufacturing, it is essential to extend their lifespans for as long as possible or ensure that these technologies are recycled and their components reused elsewhere (Jensen, 2019). Recycling turbines might lead to savings of over 7,000 tonnes of CO_2 that would otherwise be emitted through the disposal of the metals and materials contained within them (Jensen, 2019).

While wind turbines themselves have a high potential to be recycled, their blades are particularly challenging in how they might disposed of or reused. They are made of various materials, including glass, carbon or basalt or thermoplastics, polyvinyl, copper wiring, and steel bolts. All have different properties and requirements for recycling. The estimated lifespan of a wind turbine blade is between 20 and 25 years (Geiger et al, 2020). A potential 15,000 wind turbine blades may need to be removed and recycled in the next five years (ETIP Wind, 2019). Both the appropriate and safe management of these blades and the emergence of new more sustainable designs is essential to any sustainability of wind power. Wind energy companies are making progress on this. In 2022, Siemens Gamesa powered its first turbine using its RecyclableBlades, parts manufactured so that their components can be fully reclaimed at the end of their lifespans. Made of fibreglass, resin, and wood, the blades' components can later be separated through an acid solution and then reused for other products, such as the casings of flat-screen TVs (Siemens Gamesa, 2022). The disposal of these blades doesn't necessarily have to involve their being stripped of their different components, which is difficult. If they cannot be directly recycled, the blades can be repurposed for inventive, imaginative purposes, being re-used in playgrounds or bike racks. They can also create new shared urban spaces: such as skate parks, wave-breakers and pontoons at beaches, or bus shelters (Re-Wind Design, 2018).

There is a need for future energy policies to understand and treat renewable energy technologies as materials to be recycled and reused, rather than as waste to be managed (Kumar and Turner, 2020). Treating them as waste risks the enrolment of solar technologies into the same chains of cost-shifting evident in electronic waste and, potentially, smart meters. While recycling won't eliminate the need to mine new materials, it could reduce the new extraction required (IEA, 2021a) and, with it, stop these technologies from going to landfill and entering the same geographies of waste that lead to e-waste being processed in places like Agbogbloshie in Accra, Ghana. For an energy transition to be globally just, it must ensure that the lifecycles of renewable energy technologies do not end at landfill sites but remain part of an energy transition through their reuse, repurposing, or recycling.

Rule 6: Recognise that a just energy transition here must equal justice everywhere

The 'green-ness' of a renewable energy transition is not assured. Renewable energy infrastructures are associated with the appropriation and enclosure of land, the displacement of communities and the seizure of raw materials for global supply chain – leaving local communities to face impacts (Dunlap and Marin, 2022). This has led Dunlap (2021b) to coin the term 'fossil fuel +' to describe renewable energies and highlight the injustices in current energy transitions that are more commonly associated with fossil fuel energies. The cases detailed in this chapter represent a process of cost-shifting in renewable energy transitions, in which finance-rich but resource-poor countries in the Global North extract from landscapes and communities in the Global South to support their renewable energy transitions (Kolinjivadi and Kothari, 2020). Renewable energy technologies in the current model of decarbonisation are based on processes that site impacts in certain places and the benefits elsewhere. These impacts and injustices across renewable energy supply chains are cumulative, building up across multiple sites (Garvey et al, 2022). They are also embodied within the products and technologies that underpin many of the energy projects discussed in previous chapters.

Energy transitions always interact with broader patterns of climate injustice, in which the everyday lives of communities in the Global South are impacted and made more precarious by emissions in the Global North (Sultana, 2022). However, the same is true in the policies that are enacted to address and slow climate breakdown. Decarbonisation in the Global North is built upon supply chains and communities, such as those in the lithium triangle or Agbogbloshie in Accra, Ghana, who are placed at the sharp end of energy transitions. The experiences of these communities highlight that a just energy transition is as global as it is local. There is a need to reframe the policies that are currently dictating what an energy transition is. As they stand, they are currently focused on supporting our lifestyles through expanding renewables in the Global North (Ajl, 2021). Ways to move forward include both the cancellation of the debt of countries in the Global South and better, more constructive discussions about reparations, both for the historical responsibility for climate change and to provide a form of reparative justice for historical ills of colonialism and global emissions (Paul and Gebrial, 2021).[7]

The years since the 2015 Paris Agreement on climate breakdown have seen increased calls for global climate action to place objectives of climate finance at their core. The term has varying definitions but a key logic: the provision of financial support for climate change mitigation and adaptation. Numerous climate conventions, including the 1997 Kyoto Protocol and 2015 Paris Agreement, have called for increased financial assistance from richer countries to those who are less wealthy and more vulnerable to the impacts of climate breakdown. A framework of how such climate finance might be provided can be found in an announcement made in the days before the COP26 summit in Glasgow in November 2021. The governments of the European Union, the UK, and the United States issued a joint statement with South Africa that mobilised billions of dollars to help promote and finance a just transition in the latter. The declaration celebrated the start of a 20-year Just Energy Transition Partnership (JET-P) to accelerate South Africa's phasing out of coal and support those mining communities and workers affected by the process). This was an ambitious plan that pledged US$8.5 billion (£7.5 billion) of financial support to support the South African government in its aim to expand renewable energy and prevent 1.5 gigatonnes of greenhouse gas emissions over 20 years (Gov.UK, 2021a). Welcoming the JET-P in 2021, the President of South Africa, Cyril Ramaphosa labelled it as a: "Long term partnership that can serve as an appropriate model of support for climate action from developed to developing countries, recognising the importance of a just transition to a low-carbon, climate resilient society that promotes employment and livelihoods" (UK Government, 2021a). Ramaphosa explicitly adopts a language of just transition, highlighting the importance of supporting workers and communities at the sharp end of an energy transition. The South African electricity grid heavily has remained heavily dependent on coal and a creaking electricity infrastructure that had led to a severe energy crisis and rolling blackouts. As a result, the JET-P financially supports the country's transition away from fossil fuels while ensuring energy access and affordability (Bartlett, 2022). South Africa's decarbonisation also overlaps with policy priorities of supporting workers in declining fossil fuel industries. More than 92,000 workers are employed in the country's coal sector – all are at risk in an energy transition (ESI Africa, 2021).

A successful JET-P between South Africa and the donor countries will likely become a model for future climate finance instruments and how they can simultaneously reduce emissions, address energy poverty, and support workers. In 2022, G7 countries announced a similar partnership with Indonesia. Future JET-Ps must tailor their model of funding and priorities to the countries they are seeking to support and help address the issues founds. However, there cannot be a one-size-fits-all approach. In many countries, demand for electricity is skyrocketing, leading to many governments having to make trade-offs between climate action and energy security.

There is a need to recognise that nationally-driven energy transitions in countries like the UK are only one point in a global web involving the extraction of resources, the manufacturing of goods, and the disposal of waste. Decarbonisation here creates demands on resources and labour that impact people and communities elsewhere. Global energy transitions have significant local and regional impacts that must be addressed in future global climate policies. The removal of fossil fuels from one national energy grid reduces that country's emissions but can also accelerate injustices in other places like the 'lithium triangle' or the Democratic Republic of Congo. A cosmopolitan vision of climate justice and equity must be at the centre of a just energy transition. This involves acknowledging and understanding how energy transitions in our homes, neighbourhoods and energy grids will have a social and environmental impacts in other parts of the world. Those working towards a just energy transition should not only assert the importance of some communities and places but stress the need for justice in all communities across the supply chain and energy grid.[8] In using electricity in my home in Bristol, I am not divorced from communities in Bolivia, the Democratic Republic of Congo, or Ghana. All support a move to low-carbon energy through their labour and bear the often-hidden impacts of decarbonisation. A failure to include these communities and to make space for them in decarbonisation would merely repeat the injustices of the past.

9

Conclusion

As I speed along the train line between Gloucester and Birmingham in the West Midlands of the UK, I look out across the green landscape as field gives way to thicket. The Defford Radio Telescope in the near distance is turned upright to the sky. Birds fly above, silhouetted against the blue. Empty fields soon become crowded. Row upon row of solar panels stand equally spaced, all facing south and spread out across 60 acres on land previously owned by a Royal Air Force base. It is from this base that early radar systems were trialled during the Second World War. After the war, the airfield proved too small to support the new landing and taking off of new aircraft and fell out of use in the 1950s. Today, these fields give way to panel and frame. As I look out of the window, one field holds fully assembled panels ready to generate electricity. The next is home to metal frames only. Solar PV panels will soon join them. Another field lies empty, waiting for the assembly line of renewable energy to arrive. In this pocket of Worcestershire, the energy transition is taking form.

This book has not been motivated by questions of 'how we will decarbonise our energy grid?' or 'which renewable energies will be adopted?'. Those questions are mostly being answered already. Renewable energy infrastructure is an important part of the energy model: 174 GW of solar and 97.5 GW of wind energy capacity was installed globally in 2021 alone (Rai-Roche, 2022; WWEA, 2022). Decarbonisation is already taking shape in fields like those near Defford, the coastlines in south England, and homes and communities in Bristol and Glasgow. The question that we should ask is 'how can an energy transition be made fair, equitable, and just?' This question has never been more urgent, with the current energy model's inadequacies exposed in the context of rising energy prices and increased energy insecurity. An energy transition is no longer about climate action exclusively. It is about keeping the lights on with cheap, reliable and renewable energy. It is about supporting people and making their lives better.

The decarbonisation of our energy grid is a transition from one set of circumstances (burning fossil fuels for energy) to another (not burning them).

These sorts of transitions are messy, creating new geographies of winners and losers (Bridge et al, 2013). While there will likely be more winners than losers (in planetary terms, at least), there will be many negative impacts of decarbonisation. The geographies of energy transitions are uneven, evident in where new energy infrastructure will be manufactured, built, and recycled. It will also take place across our homes, neighbourhoods, and communities. Across these spaces, decarbonisation interacts with the dynamics and tensions that are already present, which may give rise to a sense of injustice or illuminate a narrowness of the support available to cope with transitions.

When injustice is felt, efforts to challenge, stop, or change policies are rarely far behind. A telling example is found in the *Gilets Jaunes* ('Yellow Vests') protests in France that began in November 2018. These weekly protests, in which people donned fluorescent high-vis vests and block roads, had roots in rising fuel prices between 2017 and 2018. While rising costs were linked to world prices and supply, they were also increased through fuel taxes on both petrol and diesel. Climate policies often adopt demand-side interventions that seek to change how people consume or use fossil fuels or other polluting products. This had led to fuel tax increases in France, with these measures seen as a way to encourage people to drive less. However, these policies involved limited discussion of the equity of such a move which, if applied in a blanket way, would unfairly punish those living in rural areas. Rural communities often have no other way to get around and drove diesel cars because they could go further on what was previously a cheaper fuel (Charmorel, 2019). Fuel tax rises also came at a time of rising anger at French President, Emmanuel Macron (2017–), directed at plans to decrease the speed limit on rural roads, blunt comments about people reliant on state welfare payments during an era of rising costs, and political scandals. For many, Macron appeared to be detached from the economic challenges of 'ordinary' people (Charmorel, 2019). The fuel tax rises became one symbol of wider anger and tension between who was 'winning' and who was 'losing' in French society.

Weekly protests took place every Saturday across cities, as well as small and medium-sized towns. On 17 November 2018, 300,000 protestors took to the streets (Grossman, 2019). The yellow vests staged roadblocks at 1,500 strategically selected roundabouts (Charmorel, 2019). Others vandalised speed cameras, shop fronts and public monuments. This action was often met with extensive police violence and global headlines. While the protestors came from various backgrounds, most lived in rural areas. Few had been part of political protests before (Charmorel, 2019). By 10 December 2018, the protests had exercised enough dissent that Macron's government cancelled further increases to fuel taxes and increased minimum salary requirements.

The yellow-vested protestors were seen as drawing attention to problems faced by many communities in France: with one poll at the time finding that 70 per cent of those surveyed supported the protests (Charmorel, 2019). They have been presented by some as a symbol of a gap between popular expectations of successive French presidents, including Macron, and their actual policies once elected (Grossman, 2019). The protests took place in the wake of the 2017 Presidential election in which Macron beat Marine Le Pen, the far-right populist leader of the *Rassemblement National* party in a run-off election that had exposed a division in the French electorate beyond traditional splits of left vs. right. Macron's voters were more likely to live in cities and be well-off, while Le Pen's voters were more rural, younger, and poorer (Charmorel, 2019).

The *Gilets Jaunes* protestors provide important lessons for a just energy transition. Their protests disputed and discredited the idea, long held by policy makers, that climate action could take place without paying attention to inequality (Kinniburgh, 2019). The yellow vests came from parts of the French population that would likely be most affected by the fuel tax rise: workers who, living in rural areas, relied upon transport for their livelihoods, or those on low incomes who would be unable to pay more for fuel due to stagnating wages. These groups will also be disproportionately impacted by any additional carbon tax or future changes to energy prices.

Many movements, including the Don't Pay Campaign and Enough is Enough, emerged as a response to the energy price crisis and the rising cost of living in the UK in 2022. These groups worked hard to highlight how the current energy model is leaving people behind, and that, for some, energy transitions don't address their needs and priorities. These campaigns – like the *Gilets Jaunes* protestors in France - highlight an important tension of decarbonisation in an era of rising precarity and economic inequality: between the narratives of environmentalists anxious about the 'end of the world' and those who are concerned about making it to the 'end of the month' (Martin and Islar, 2021).

While the urgency required for climate action means that we need to act now, this can lead to policy overlooking important complexities linked to fairness, equity, and justice. The focus can often be on the transition from fossil fuels to renewables– not on what might make such a process just. This creates space for friction and protest, like the outpouring of anger by the *Gilets Jaunes*. This is also evident in the words of Gary Smith, the general secretary of the GMB trade union, in an interview in 2022:

> We should not get caught up in a bourgeois environmental debate driven by the bourgeois environmental lobby … The debate on the left needs to seriously talk about climate change, but it needs to be focused on jobs. And the renewables industry, and many of those who espouse it in politics, have no interest in jobs for working class communities. (Wearmouth, 2022)

Suspicion of energy transitions has also provided a space for politicians to highlight the perceived 'unfairness' of decarbonisation and use it to their advantage. The cost-of-living crisis in the UK has been accompanied by increasing calls to abandon government ambitions to decarbonise. Those who pushed for Brexit in the UK have now levelled their sights on energy transitions (Horton, 2022; Horton and Taylor, 2022). This represents a pivoting in right-wing narratives of climate change. They no longer dispute climate science. Instead, they are arguing that decarbonisation is occurring too fast, too widely, and paying little attention to the impacts it has on people, households, and communities (Atkins, 2022). These politicians link renewables to rising energy prices, painting energy transitions as undemocratic and driven by 'elites' and of not solving problems that 'normal' people face.

Those working for an energy transition need to dispute these narratives. Opposition to decarbonisation is not widespread: seven in ten people in the UK polled in 2022 saw the expansion of renewable energies as a solution to the cost-of-living crisis (Cuff, 2022). This support can be further encouraged by ensuring that energy transitions are socially good as well as leading to reduced emissions. People are more likely to support climate policy if it comes with wider pledges of social and economic reforms, such as wage increases or the expansion of affordable housing (Bergquist et al, 2020). Decarbonisation presents an opportunity to do things differently and to re-write what the energy sector is, what it looks like, and how we are linked to it.

Popular support for renewable energy investment is high. In the UK in 2021, 79 per cent of respondents to one survey voiced support for renewable energy (BEIS, 2021b). Yet, despite this support, renewable energy transitions in the UK remain characterised by missed opportunities. In the UK, solar farms have been refused planning permission at dramatic rates, with permission for 23 projects refused in England, Wales and Scotland between January 2021 and July 2022 (Horton, 2022). This was an increase from previous years: only four projects were refused between 2017 and 2020. In the summer of 2022, the short-lived government of Prime Minister Liz Truss argued that solar farms on agricultural land were at odds with goals of energy and food security (Craig, 2022). Truss wasn't an outlier: in the leadership election that placed her in 10 Downing Street, her defeated opponent and subsequent successor in government, Rishi Sunak, voiced agreement with such claims: arguing that on his watch, solar farms would not be built on 'high-quality' farmland (Martin, 2022). In the House of Commons, Conservative MPs have campaigned against solar projects in their constituencies, arguing that they are turning rural areas into industrial landscapes, and suggesting that investments should be made in rooftop solar instead (Gabbatiss et al, 2022).

Yet, many people remain priced out of the retrofitting measures and rooftop solar that can reduce both household emissions and energy bills. While I have written about rooftop solar, I think it's important to stress that my roof remains bare. In late 2021, we got a quote for rooftop solar on our home via the Solar Together scheme discussed in Chapter 6. For eight solar panels, including installation, we were quoted £4,173. The prices increased by £2,228 when additional battery storage was included. This is a large sum for any household, particularly when facing other costs. Schemes like Solar Together are worthy of praise: they allow people to club together to reduce the costs of decarbonising their homes. However, they are out of the reach of many who can't afford to put up that amount of money. Many people have less than £500 in their savings (Jack, 2022). Many more are facing stagnating wages and spiralling prices. Others do not own their homes. A reliance on schemes that focus upon people investing their own money in household energy transitions neglects that such opportunities are not accessible to everybody. More must be done to help those most vulnerable to energy poverty.

Missed opportunities can also be found at the local and regional level, where local authorities are struggling to decarbonise and support their communities in doing so. While 75 per cent of local authorities in the UK have declared a 'climate emergency', their climate plans often remain ambiguous and vague and few authorities have involved their communities in decision-making (Gudde et al, 2021). The opportunities missed by local authorities in the UK can, at times, be blatant. In 2022, Medway Council (a local authority in Kent) rejected its planning application to install solar panels on the roof of its own office building.[1] This rejection was because the proposed panels would introduce a 'modern, incongruous, and out of character feature' to the building (Hunter, 2022). Medway Council had declared a climate emergency in April 2019.

In 2022, local governments stepped in to help residents during cold weather and high household bills. Many opened 'warm banks', publicly accessible spaces where people could sit in comfort and warmth. However, the capacity of local authorities to go beyond these measures is limited. Many councils in the UK have explored the possibilities of retrofitting housing stock and supporting others in doing so, yet they remain unable to due to the costs of such action and the need to fund decent public services elsewhere. Funding for local governments in the UK has been squeezed for over a decade, severely limiting the capacity of any authority to pursue decarbonisation. Funding was cut by 49.1 per cent in real terms between 2010 and 2018 (UK Parliament, 2019). There is only so much capacity at the local level, where decisions are constrained by national governance and energy policies.

National change is required. The energy price crisis in 2022 has its roots in decades of national energy policy making. Global turbulence has caused a crisis felt across the globe — with global drivers including a prolonged winter and stronger demand for gas from Latin American and Asian economies. Nationally, the UK faced challenges caused by a fire on a subsea cable carrying electricity from France to Kent and slower-than-usual wind speeds in late 2021 (Buli and Jacobsen, 2021; Sheppard and Wilson, 2021). The convergence of these challenges with the crisis caused by the Russian invasion of Ukraine exposed vulnerabilities inherent in the national energy model. For several decades, successive governments have maintained a reliance on gas to both power our homes and keep them warm. The UK became a net importer of gas in 2004, meaning it was reliant on supplies from elsewhere. However, despite this reliance, the UK has a lack of infrastructure to store gas. The Rough facility, housing gas under the North Sea off the East Yorkshire coast, was closed in 2017 by its operator, Centrica. The facility accounted for around two-thirds of national gas storage capacity and was not replaced. In 2021, while France had 14 weeks' worth of gas reserves to protect against shortages, the UK had an estimated four days (The Spectator, 2021). While the Rough storage facility was reopened in late 2022, the vulnerability of many to high energy prices remains. Some 78 per cent of UK households were reliant on gas central heating to keep warm (BEIS, 2022a). Without concerted action, all remain at the mercy of global energy prices and flows.

Policies that might have addressed this reliance and stimulated the decarbonisation of homes previously fell victim to Conservative austerity policies and Prime Minister David Cameron's 2013 call to 'cut the green crap'– a cynical reversal from previous images of him hugging huskies and calling for the electorate to 'vote blue, get green' (Groves, 2013). Total energy bills in the UK are estimated to be £2.5 billion higher than they would have been without such a decision (Evans, 2022). Cameron's focus was on the short-term, on restricting government spending to reduce budget deficits. As a result, his government missed key opportunities to both reduce household emissions and protect many from future spikes in energy prices.

Government policies to address the energy price crisis in 2022 missed key opportunities. That year, three successive Conservative governments (led by Prime Ministers Boris Johnson, Liz Truss, and Rishi Sunak respectively) introduced policies to address issues of rising energy prices and the cost of living. All represented a similar vein of short-term thinking, with little change made to an energy model that had led to the vulnerability of many. The first government, led by Prime Minister Boris Johnson (2019—2022) announced a series of payments to support households facing spiralling energy bills in May 2022. The then Chancellor of the Exchequer, Rishi Sunak, announced £400 discounts on energy bills provided to all households (in the form of a loan to be repaid in several years' time), with an extra £650 made available to those who receive social welfare payments (Swinford, 2022). There were issues with these payments. While energy companies directly applied this discount to the energy bills of many, some struggled to access this support. Those on pre-payment meters had to be sent vouchers, which they would then add to their pay-as-you-go account. Many reported not receiving vouchers and, as a result, being unable to pay spiralling bills (Wearn and Masud, 2022).

This package of financial support was paid for by an announced windfall tax on the profits of energy firms. Energy companies in the UK were to pay an extra 25 per cent on their profits, in addition to the 40 per cent they were already liable to pay — this was expected to raise close to £5 billion to finance the support given to households (Mayes, 2022). Such a move followed loud calls by many to tax the profits of many companies that hit record highs in 2022. For example, Shell's profits in the first four months of 2022 hit US$9.1 billion (£7.4 billion) (Shell, 2022). However, this windfall tax was accompanied by another policy, in the which the UK government announced a new plan for tax relief for new oil and gas projects, with 91 per cent of every £1 spent on 'UK extraction' covered by tax savings (George and Mace, 2022). New renewable energy projects were not included in this allowance. While the government refused to call it a 'subsidy', this new tax relief represented taxpayer-funded financial support for fossil fuels that outweighed any tax on profits.[2]

Prime Minister Boris Johnson resigned from office in July 2022, after a series of political scandals. A key policy challenge faced by Liz Truss's new government was the continued rise of energy prices. In August 2022, Ofgem announced that it would raise its energy price cap, with the average UK household gas and electricity bill increasing to £3,549 a year. This was close to treble the average bill of only a year before (the cap in October 2021 was £1,277). To address these energy prices, Truss's government announced an Energy Price Guarantee that would cap the average household energy bill at around £2,500 per year. A different scheme was set up to support businesses, likely reducing their bills by thousands of pounds. At the time they were introduced, the support packages were estimated to have cost the government £150 billion (Parker et al, 2022).

The Energy Price Guarantee gave people some respite from increasing prices over the winter months of 2022. However, it did not represent a proactive attempt to solve the entrenched problems that caused energy prices to be high. The focus of both interventions was on keeping the lights on today, not on a transition towards energy security and very little on creating a future energy system that is *just*. People with second homes were also able to receive the £400 discount on energy bills, while renters who paid their bills as part of a monthly payment were at risk of missing out (Godfrey, 2022; Tapsfield and Heffer, 2022). Truss's support package was accompanied by a suite of tax cuts that would have benefited the richest households substantially more than those on low incomes, part of a series of measures announced by Chancellor of the Exchequer Kwasi Kwarteng in October 2022. The energy support introduced was overshadowed by these changes, due to their impact on global market confidence in the UK which led to a rapid fall in the value of the pound and, ultimately, the end of Truss's 44 days in office.

Truss was replaced as Prime Minister by Rishi Sunak in October 2022. To mitigate some of the damage caused by Truss and Kwarteng's budget, a new Chancellor, Jeremy Hunt, announced new measures in November 2022: including tax rises and spending cuts. Within these measures was the extension of Sunak's previous windfall tax to include low-carbon energy generators, such as those generating wind, solar and nuclear energy, and an increase in the levy paid by North Sea oil and gas companies. This new levy on renewable energy generation targeted those who generated more than 100 GWh a year and, as a result, did not affect smaller, community-driven energy projects. Hunt also announced a doubling of annual investment in energy efficiency measures and the setting up of an 'Energy Efficiency Taskforce' (formally established in March 2023). However, funding for retrofitting was delayed to after 2025. In short, Hunt promised future spending rather than introducing measures that would help people immediately. There would be a three-year wait (and a general election) before any investment.

No policy introduced in 2022 by the Johnson, Truss, and Sunak governments included a wholesale focus on household energy efficiency and retrofitting to reduce energy bills as soon as possible.[3] Government spending went elsewhere. The Johnson government's tax relief for new oil and gas projects, if redirected, could have been used to insulate 2 million homes (Graham, 2022). Without including energy efficiency policies, the solutions represented a circular exchange of money – from the energy company to the government (via the windfall tax), the government to the consumer (via support payments), and the consumer back to the energy company (via paying bills). Nothing was done to address the failings of the energy model. Even with a new windfall tax, energy companies still walked away with vast profits paid for by spiralling energy bills. In November 2022, the chief executive of Care England accused gas suppliers of profiteering by forcing care homes to pay 'horrendous and financially crippling rates' (Lawson, 2022). These are common stories. Some schools, for example, faced price rises of over 500% in the autumn of 2022 (Belger, 2022). The policies of 2022 were quick fixes only. The status quo was left unchanged. Another opportunity for a just energy transition was missed.

Rules for a just energy transition

Political leaders in the UK have sought to present the country as having the potential to be a renewable energy leader on the global stage. Before the COP27 summit in Sharm El-Sheik, Egypt in November 2022, Prime Minister Rishi Sunak told the UK House of Commons: "We will make this country a clean energy superpower. We will accelerate our transition to renewables which have already grown four-fold as a proportion of our electricity supply over the last decade" (Prime Minister's Office, 2022). The day before, the leader of the Labour Party, Sir Keir Starmer, labelled Sunak a 'fossil fuel prime minister' yet adopted an identical rhetoric, arguing that 'Under my Labour government, the UK will become a clean energy superpower.' The focus of many energy policies in the UK has remained on expanding supply. At the time of writing, Nationally Significant Infrastructure Projects in the UK that are under consideration include solar farms near West Burton on the Yorkshire Dales, Aldington in Kent, and Gainsborough in Lincolnshire. New offshore wind projects are spread across the coastline, from the Five Estuaries project (37 km off the Suffolk Coast) to the Mona Offshore Wind Farm in the Irish Sea. In 2022, the Labour Party pledged to expand supply if voted into government. This included policies to double onshore wind capacity, triple the amount of solar energy, and more than quadruple offshore wind power in the UK – all by 2030. In late 2022, a group of Conservative MPs also lobbied for an expansion of onshore wind, rebelling against government policy in seeking to remove planning restrictions on the technology. In an era of a national energy crisis, renewable energies remained widely discussed but politically divisive.

In being deemed of 'national significance', large renewable energy projects are ultimately approved by the national government, rather than local authorities. All these projects signal the continued tension between the national benefits and local impacts in an energy transition. While national benefits might include emissions reductions and energy security, local communities face a changing landscape and the disruption and impacts caused by project construction. If the UK is to become a 'renewable energy leader' at the global level, it needs to ensure that its energy transition is inclusive and just. Addressing this tension is one step towards doing so.

Across this book, I have sought to illuminate how an energy transition that is nationally led but community-centred can create a new energy model. A just energy transition requires not only the reduction of emissions but includes helping vulnerable households and providing new futures for communities across the electricity grid. In outlining this vision, I have detailed six rules for such a transition. These are:

Rule 1: Push for community-centred renewables: Climate action requires expansive action to reduce greenhouse gas emissions. The scale of action required often leads to projects being large and expanding across landscapes in rural areas. This results in the creation of energy peripheries in which communities, many of whom have been previously neglected by policy, see their landscape forever altered to provide 'green' electricity for those elsewhere. However, renewable energies can come in various sizes and, as a result, a just energy transition can include smaller facilities that support local communities directly and become tied to the places where they are built. While these facilities might spring from the ground up, they require extensive financial and technical support. Low-interest grants and loans must be made available to communities seeking to invest in their own energy infrastructure and new spaces created for communities to learn from one another and develop their own place-based approach to energy justice.

Rule 2: Elevate and emphasise the participation and voices of communities when developing new energy projects: A just energy transition is about more than how the benefits and burdens of the energy model are distributed. It must also involve an affirmation of the right of people and communities to be heard and listened to in energy decision-making. As I detailed in Chapter 4, local communities must be treated as having a greater role than just hosting renewable energy infrastructure on their doorsteps. The current energy model often prioritises shallow forms of consultation and compensation to appease local communities. This can be helpful, providing new facilities and forms of support to residents. However, it can also perpetuate broader patterns of injustice, such as rewarding landowners with financial compensation but not addressing the broader impacts on the landscapes. In a just energy transition, people and communities can be reconnected to renewable energy infrastructure through projects that they influence, own, and benefit from.

Rule 3: Foreground community-centred energy schemes in local economies and wealth building: While climate action requires a concerted effort from the national government, schemes that prioritise community participation, address residents' needs, and keep financial benefits in the immediate area represent a core part of a just energy transition. The income gained from directly owning a wind turbine can be 34 times the amount that these communities receive in support or benefit payments from private energy companies (Aquatera, 2021). A just energy transition would empower people and communities to use these funds in ways that address their needs. Energy infrastructure and policy can play an important part in processes of community wealth building, in which local economies become democratised and transformed by active intervention. Examples from both Preston in Lancashire and North Ayrshire in Scotland highlight how regional organisations, such as local authorities and large employers, can adopt policies that focus on boosting local economies and creating jobs by investing in renewable energy generation. The use of Community Municipal Investments represents an opportunity for local energy schemes to become embedded within the community and local economies, linking residents and their investments to new energy infrastructure and a low-emission electricity grid.

Rule 4: Prioritise those most vulnerable to energy poverty: Spiking energy prices have illuminated the vulnerability of many and led to calls for mass retrofitting schemes. Emergent plans for energy efficiency measures and the expansion of rooftop solar must place restorative justice at their heart. Energy poverty and its knock-on consequences have particular geographies, affecting some households and communities more than others. This vulnerability is not abstract, merely formed of numbers on a spreadsheet. Instead, it is shaped by embodied and everyday experiences of cold, damp, and dark homes and, often, an inability to do anything about it. Future energy policies must prioritise those who are disproportionately vulnerable to price spikes and energy poverty (such as those on low incomes, or those with disabilities or chronic health problems) and those who do not have the resources or autonomy needed to make the changes required (those in flats, rental properties, or social housing). Lastly, the geography and architecture of energy efficiency interventions in the UK should be targeted to neighbourhoods and communities where vulnerability is clustered, such as inter-war housing estates. Financial support for these households might be given through 'pay as you save' tariffs, or accommodating lease agreements, themselves linked to local community wealth building, job creation, and green skills policies.

Rule 5: Ensure the participation and inclusion of workers: The urgency of decarbonisation has led to a situation in which short-term gains (of lower emissions) have been prioritised over the provision of better livelihoods and working conditions for those whose labour supports such a transition. Energy transitions should prioritise notions of good, green work and working conditions through the adoption of a longer-term vision. This should include a focus on green skills, providing workers with the training and opportunities necessary to not only transition from one sector to another but to see their future in the renewable energy sector. Schemes should be targeted to address current issues of inclusion and diversity. These steps have positive consequences, addressing problems of gender blindness in policy, investing in young people's futures, and providing better jobs in local economies. Worker-led action, both in response to the precarity of their work and in calling for new green jobs, demonstrates the significance of workers in an energy transition, and how they hold a vision of what their role might be. These voices must be heard, elevated, and placed at the centre of a just energy transition.

Rule 6: Recognise that a just energy transition here must equal justice everywhere: Future policies and schemes must avoid a cost-shifting of impacts to countries, regions, and communities elsewhere. This includes the impacts linked to the abstraction of resources, such as lithium from the salt flats of Argentina, Bolivia, and Chile and cobalt from the Democratic Republic of Congo. Energy transitions have also been implicated in forced labour in Xinjiang, China, toxic working conditions faced in Ghana, and repressive politics in Western Sahara. The experiences of these communities illustrate how a just energy transition is as global as it is local and the need to reframe how we understand decarbonisation. Current transitions are focused on a status quo of unequal exchange of impacts and benefits between the Global North and Global South. New policy frameworks are required to rebalance this exchange and provide concerted climate finance to those countries facing impacts, both from climate breakdown and the supply chains that are central to mitigating it. Just Energy Transition Partnerships are one step, but they should be scaled up to include more countries and provide context-specific and broader accommodating financial support. Such moves would further tie together the experiences of energy transitions in one part of the world with communities elsewhere, albeit perhaps in a more positive way.

Energy transitions are dynamic: new technologies will emerge, and others will be made redundant. This book will be out of date in a few years — and that won't be a bad thing. Technologies such as vast utility-scale batteries, hydrogen, or pumped storage hydropower would create flexibility in a future electricity grid. This could be via extending the availability of solar and wind energy (through storage in batteries). Elsewhere new sources of energy might be pioneered and introduced (such as hydrogen). In the home, domestic energy demand can also be made more flexible and efficient through the shifting to new technologies and ways of managing electricity. A future energy model might include localised energy markets that are designed to meet time- and place-specific demands of the national grid by using data collected by smart meters. These ideas are already being trialled. In Oxford, there is a pilot project of residents trading electricity between them, with energy flowing from home to home as they have their different moments of peak demand. In Brixton, London, a block of flats has become a hyper-local energy market, with solar on top of the building stored in a battery and sold, bought and traded by residents. At a wider level, electricity prices can become variable based on where you live: with those living close to energy infrastructure able to pay less. It is estimated that variable local electricity prices might save a total of £35 billion on their bills by 2035, with those living in northern England and Scotland most likely to benefit (Gosden, 2022).

There is also a need to think beyond electricity. Household energy transitions must now pay more attention to heating and cooling. Burning oil and gas to heat both rooms and hot water in our homes is the biggest source of buildings emissions in the UK (CCC, 2022). The roll-out of heat pump technology would help address this — but such a process remains complicated by industry lobbying (Webster, 2022). The summer of 2022 highlighted how many homes and buildings in the UK are not equipped to deal with heat. Temperatures hitting 40°C exposed the need for expansive discussions about how we can keep our homes and workplaces cool and support those vulnerable to heat stress.

Within a just energy transition, small-scale, local-led or community-centred renewables can help address issues of energy poverty and give local communities an element of control over their own energy transition. Larger projects can create thousands of jobs and generate new opportunities for skills and retraining. Both can be linked to policies of community wealth building, in which the financial benefits of new energy schemes flow within the local community itself. This would represent an energy transition that builds new forms of popular support, while challenging those who seek to discredit net-zero agendas. Energy transitions need to be further broadened to address social and economic questions, as well as technical or scientific ones. They can also be deepened to better include people and communities and their circumstances, geographies, and complexities — rather than primarily focusing on global and national narratives of reducing greenhouse gas emissions. This is a political opportunity to build support for climate action and mobilise communities in developing a new energy model (Aronoff et al, 2019). Energy generation can be re-localised and reconnected with communities and local economies. New lines of secure, good work can be created. Homes and housing improved, and people elevated from energy poverty. Lives can be made better as electricity is put to work for our benefit. We can have a say and a stake. For any energy transition to be *just*, these opportunities must be taken, and they must be taken now.

Notes

Chapter 1

[1] According to the Energy Price Index, in September 2022, electricity in Portugal was charged at a unit price of €0.25 per kilowatt hour (22p). In the UK, the equivalent price was €0.39 cents (35p).

[2] All currency conversions are for illustrative purposes and are based on exchange rates in early December 2022.

Chapter 2

[1] Understood as an area of land that could be ploughed by a 'team' of eight oxen in a year, estimated at 120 acres.

[2] Data taken from Open Domesday project – https://opendomesday.org/place/TR0964/seasalter/ (Accessed 24/01/2022).

[3] Each bank, although of varying sizes, will likely have as many panels as possible – with panels organised in an east-to-west orientation to ensure effective land use. Panels will be raised at different heights – from 2.6m to 4m above the ground (both to avoid flooding and allow sheep to graze underneath them).

[4] Information on opposition to 'Project Fortress' (formerly known as Cleve Hill Solar Park) is taken from submissions to the UK National Infrastructure Planning (2021) portal during the planning permission process – namely the submissions with the reference numbers: AS-013, AS-016, REP7-090, REP2-073, and REP5-051.

[5] This violence is traumatic. On 21 June 2020, 15 people were murdered at Huazantlán del Rio, allegedly due to disputed elections (Ferri, 2020).

[6] The use of the term 'Superfund' alluded to the federal plan – part of the 1980 Comprehensive Environmental Response, Compensation and Liability Act – to commit funds to clean up contaminated and polluted lands (Just Transition Research Collaborative, 2018).

[7] The International Trade Unions Congress (2020) keep a series of scorecards that keep track of the presence of just transition and social dialogue policies in state government's Nationally Determined Contributions, submitted to UNFCCC Conferences of Parties. At the time of writing, only nine countries and the European Union have such policies.

[8] Erin Savage of Appalachian Voices in a discussion (21/10/2021).

[9] Email from the Energy and Climate Change Directorate of the Scottish Government, 08/07/2021.

[10] Ryan Morrison of Friends of the Earth Scotland (06/09/2021).

[11] Taken from submissions to the UK Planning Inspectorate during the planning permission process – namely AS-059.

Chapter 3

[1] All members of Bristol Energy Cooperative (including myself) receive interest on investments, generated from the money made on the energy produced.

[2] My thanks to Will Houghton at Bristol Energy Cooperative for discussing this process and BEC's experience of it with me (11/02/2022).

[3] Across this book, I use the term 'Black Communities' to refer to those who have a shared history of European colonisation, imperialism, ethnocentrism, neo-colonialism, and overt and structural racism. The use of the word 'communities' in the plural highlights the variety of groups who share this history, including (but not limited to) African, Caribbean, South Asian, Latinx, and Indigenous communities. When a particular source uses a certain label to describe communities, I replicate that to signal the particular focus of the evidence presented or argument made.

[4] Today's hydroelectric dams are classified by their generation potential – or how much electricity their turbines can produce in a certain period. While there is not necessarily an agreed-upon definition and differentiation, I follow others in understanding the different sizes as follows (BHA, nd; Breeze, 2019): Pico – less than 5 kilowatts (kW); Micro – 1 kW to 100 kW; Mini – 100 kW to 1 megawatt (MW); Small – 1 MW to 10 to 30 MW (the upper limit differs from country to country); and Large – >10 to 30 MW (depending on the upper limit of hydro deemed 'small').

[5] Maureen Harris of International Rivers (in a response to my questions on 23/09/2021) explained a series of misgivings about the 'sustainability credentials' of hydropower. First, the complexity of direct, indirect and cumulative impacts (and how they are linked) is often far too difficult to include in the parameters adopted in the HSAP. Instead, this complexity is flattened and narrowed, neglecting important consequences. Second, the HSAP is limited in taking into account the long-term impacts of hydropower.

[6] For more information on the environmental impacts of hydropower, please see: Nilsson et al, 2005; Agostinho et al, 2008; Ziv et al, 2012; Araújo and Wang, 2015; Benchimol and Peres, 2015; Lees et al, 2016; Fearnside, 1995; 2016; Soukhaphon et al, 2021.

[7] Annex I countries have a set number of emissions certificates, based on national reduction targets, to spend. These countries can purchase Certified Emission Reduction units (CERs) – the revenues of which would fund projects elsewhere, such as for renewable energy. This might be through investment in funding a new solar array, energy efficiency schemes or work towards rural electrification. This is based on two assumptions – first, that it doesn't matter where GHGs emissions are made or reduced and, second, that reducing emissions in Annex II countries is cheaper than in Annex I countries and overlaps with policies of sustainable development (Erlewein, 2018).

[8] Additional reasons for this gap are over-ambitious planning and the complicated planning and approval system in Bosnia and Herzegovina, – spread across municipalities, cantons, commissions, ministries and government agencies and requires 50 different types of permits (Dogmus and Nielsen, 2019; 2020).

[9] I am thankful to the interviewees from *Eko Pan* (23/09/2021) and *Centar Za Zivotnu Sredinu* (14/10/21) for discussing these issues with me.

[10] My thanks to Saskya Huggin of the Low Carbon Hub in Oxfordshire, UK for discussing the Osney Hydro project, as well the numerous community-centred renewable energy projects in the region (20/10/2021). The Osney Lock project was led by residents with Low Carbon Hub providing help towards legal costs, outreach and expertise.

[11] Christian Poirier, Amazon Watch (07/10/2021).

[12] The European Marine Energy Centre in Orkney is jointly funded by the European Union, UK and Scottish governments and the Orkney Islands Council (OIC), further highlighting the collaboration across groups and funders in Orkney's energy transition.

[13] My thanks to Rebecca Ford (23/09/2021) and Jack Breckenridge, Caron Oag and Lara Santos at the European Marine Energy Centre (01/10/2021) for sharing their thoughts and input on these discussions of Orkney's energy transition . Rebecca completed a PhD thesis at the Institute for Northern Studies, University of the Highlands and Islands in 2022. This work detailed the importance of narratives and place-based stories in marine renewables in Orkney. At the time of writing, Lara Santos is completing a PhD at the University of Edinburgh and the European Marine Energy Centre on the justice dimensions of emerging renewable energy technologies.

[14] In Orkney's case, wind energy has also become an important route to defining and redefining the island's relationship with the mainland (Dudley, 2020).

[15] Caron Oag, European Marine Energy Centre (01/10/2021).

[16] No real modifications will be made to the weir and fish passes will be built alongside the scheme that allow for fish (such as eel, sea trout, or lamprey) to continue their journeys upstream or downstream unhindered.

[17] My thanks to Jeremy Thorp at *Ynni Teg*, a Welsh energy cooperative, for highlighting the importance of community funds for many cooperatives (14/09/2021).

Chapter 4

[1] I am grateful to Rebecca Windemer for talking me through the details of what this government decision meant and still means to onshore wind in the UK (02/02/2022).

[2] Robert Kennedy, Jr is now better known for his anti-vaccine positions – he is the founder and chairman of Children's Health Defense, an activist group that has been reported as a key source of vaccine misinformation (Mnookin, 2017).

[3] In their survey, Firestone et al (2009) did not directly seek to elicit opinions of support for or opposition to the Cape Wind project. Instead, they asked about a private developer proposing the construction of 130 turbines, standing at 423 feet high, in Nantucket Sound. The words 'Cape Wind' were not included in the description.

[4] The claim of damage to spiritual heritage has been disputed (see Sheppard, 2010).

[5] I am grateful to Audra Parker of Save our Sound for discussing the opposition to the Cape Wind project with me. Audra was quick to dispute the characterisation of this opposition as 'elite' and 'NIMBY' – detailing the ways that Save our Sound represented a coalition between different groups with numerous concerns (21/09/2021).

[6] Elements of procedural justice can also be found in international law. The United Nations Economic Commission for Europe (UNECE) Convention on Access to Information, Public Participation in Decision-Making and Access to Justice in Environmental Matters (or the 'Århus Convention') was adopted in Århus in 1988.

[7] These core principles have been developed using work by Hunold and Young, 1998; Gross 2007; Simcock, 2016; and Bell and Carrick, 2018.

[8] Thrive Renewables is a renewable energy company in the UK founded in 1994 by Triodos, an ethical banking organisation based in the Netherlands. The organisation has close to 5,500 shareholders, ranging from individuals donating small amounts to pension funds. Thrive Renewables is not a project developer, so it becomes involved in a wind farm often once after the planning process and after a connection to the National Grid has been approved and made.

[9] My thanks to Adrian Warman from Thrive Renewables for talking me through how this community benefit scheme works (01/03/2022).

[10] The Rampion 2 consultation area on the UK mainland includes all or part of the district areas of Adur (population: 64,500), Arun (164,800), Brighton & Hove (277,200), Chichester (124,100), Horsham (146,800), Lewes, Mid Sussex (152,600), Wealden (160,100) and Worthing (111,400). Population figures have been taken from 2021 census data.

[11]   This ownership has historically been accepted as custom, if informal, and offshore oil and gas extraction has been managed by the UK Government (Garside, 2021). However, the 2004 Energy Act lists energy extraction rights from wind and water (such as tidal) as lying with the Crown. This domain expands out by 12 nautical miles beyond the continental shelves to halfway across the North Sea to the east, and halfway across the Irish sea to the West.

[12]   All workers involved in the transition (such as those installing solar panels) are trained via a formalised course, with these professionals then working to inform and discuss changes, costs, and benefits with residents (Mundaca et al, 2018).

[13]   Malene Lunden, *Samsø Energiakademiet* (13/09/2021).

[14]   This example of residents developing a connection to nearby wind turbines was described to me by Rebecca Ford (23/09/2021).

[15]   Described by Helen Castle of Rousay Eglisay & Wyre Development Trust, in response to questions sent via email (03/02/2022).

Chapter 5

[1]   Memories of a childhood at St Bede's were generously shared by Monica Cox, who now works at the School of Geographical Sciences, University of Bristol.

[2]   My thanks to Mark Pepper, who grew up in Lawrence Weston, is a founding member of Ambition Lawrence Weston and a key driver of Ambition Community Energy, for his time and insight (13/04/2022).

[3]   Lawrence Weston is also home to two solar PV farms (see Lacey-Barnacle, 2020). One project is owned and operated by Bristol Energy Cooperative (discussed in Chapter 3). Ambition Lawrence Weston agreed to the scheme, in exchange for 50 per cent of the profits being returned to the local community.

[4]   The challenge to the planning application came from SSE, the owner of the Seabank power station at nearby Hallen Marsh. SSE reported concerns that the wind turbine might blow over and land on (and destroy) part of its facility (not the power station itself). The challenge was submitted 24 hours before the plan went to the Planning Committee.

[5]   Olimpiu Rob of Enercoop (02/11/2021).

[6]   In 1997, 8600 residents and the municipal government of nearby Copenhagen jointly invested in the Middelgrunden offshore wind project – the only offshore wind cooperative wind farms, with a generation capacity of 10.2 MW of electricity (Capellán-Pérez et al, 2018).

[7]   Renewables in Orkney often generate more electricity than the population demands. This places pressure on transmission infrastructure – creating an issue called 'curtailment', where the electricity generated overloads the transmission cable to the mainland. This leads to energy generators slowing the turbines, resulting in a loss of revenues. While a new transmission cable remains a possibility, Ofgem (the UK energy regulator) has made any change conditional on the installation of 135 MW of generation capacity (Ofgem, 2019). To overcome this, organisations in Orkney have turned to hydrogen to store electricity generated – with it later put to use in other areas and for other purposes. The ferries that connect the islands will soon be powered by hydrogen, the first hydrogen-power aircraft flew in Orkney in 2021, and the number of electric vehicles on the islands is expanding. This provides a direct route to putting surplus electricity to direct use. There is a nice myth around this turn to hydrogen, in which the Head of Community Energy Scotland and the Managing Director of EMEC met at Kirkwall airport and developed new projects that would create new forms of demand to address curtailment – and thought of hydrogen (Ford, 2022).

[8]   The 2003 government white paper, *Our Energy Future* presented decentralised energy as important to both future energy supply and later government policy mandated ministers to promote community-led energy projects (Eadson and Foden, 2019).

[9] Aaron Priest of Viking Energy (10/09/2021).

[10] Due to space, I have treated the cases of re-municipalisation in Berlin and Hamburg briefly. For more detail, please see, Becker, 2017; Pohlmann and Colell, 2020 and Cumbers and Paul, 2022.

[11] If you want to learn more about the case of Bristol Energy, the cooperatively run *Bristol Cable* has provided exemplary analysis. A good starting point is an editorial by Old Sparky (2020).

Chapter 6

[1] Previous policies in the UK have sought to help those living in energy poverty, intervening and supporting household insulation and draught proofing. The Warm Front policy provided grants to 2.3 million households living in energy poverty between 2000 and 2013 (Putnam and Brown, 2021). This programme helped these households access energy efficiency measures – upgrading their homes, cutting bills, and reducing greenhouse gas emissions in the process. Each household gained more than £1,800 on average in saved income (Sovacool, 2015).

[2] The relative success of rooftop solar in California has led energy utilities and labour unions to lobby state leaders to restrict the incentives for the technology through the state's net metering law, with rooftop solar providing an alternative to many households from how the energy network has primarily been managed (Penn, 2020). In the US, net metering policies pay homeowners for electricity, in similar ways to Feed-in Tariffs in the UK –although US customers are not compensated if they sell more electricity to the grid than they use. However, as more and more households get rooftop solar, the number of customers paying energy companies falls, leading to these big companies losing revenue. This has led many energy companies to challenge net metering policies and attempt to disrupt energy transitions (Mulvaney, 2022).

[3] My thanks to the team at the Voices of Power project for taking the time to talk me through their work with communities, its benefits, and challenges. It's an inspirational project that deserves success and sustained support to ensure it (28/10/2021).

[4] My thanks to Andy Rolfe at the Schools Energy Cooperative for discussing this important work (17/09/2021).

[5] This point on the availability of commercial roof space in the UK is taken from an April 2022 tweet by Rosie Pearson (2022), chair of Community Planning Alliance.

[6] My thanks to Ben Delman of Solar United Neighbours for talking through this point with me, as well as the broader role of decentralised solar energy in an energy transition (31/05/2022).

[7] Thanks to Matt Wood of Energiesprung for sharing his thoughts on household energy efficiency, retrofitting in a time of rising energy prices, and government policy (16/06/2022).

Chapter 7

[1] EDF Energy is planning to expand offshore wind operations, using floating technologies.

[2] This possibility was raised by Ewan Gibbs (07/07/22).

[3] Comment submitted to Planning Office in support of BritishVolt site, 23 April 2021. Accessed via Northumberland County Council planning portal, Reference: 21/00818/FULES.

[4] This point was raised by Rebekah Diski in a discussion on 21/10/2021. I am incredibly grateful to Rebekah for her thoughts and expertise when thinking through these arguments.

[5] Rebekah Diski (21/10/21).

[6] Jake Molloy of the National Union of Rail, Maritime and Transport Workers in a discussion (21/10/2021).

[7] My thanks to Davina Ngei of GWNET (26/07/2021), Silvia Sartori of ENERGIA (02/08/2021) and Annette Hollas and Zindzi Makinde of C3E International (28/07/2021) for their generosity in discussing the issues faced by women working in renewable energy, as well as the myriad ways to address such challenges.

[8] I am grateful to Davina Ngei of the Global Women's Network for the Energy Transition for our discussions on this point. (26/07/2021).

[9] Jake Molloy of the National Union of Rail, Maritime and Transport Workers in a discussion (21/10/2021).

[10] Annette Hollas and Zindzi Makinde, C3E International (28/07/2021).

[11] I greatly appreciate my discussions with Rebekah Diski around the topic of the Lucas Plan and its broader significance today (21/10/2021).

[12] At the 2018 UNFCCC COP24 in Katowice, Poland, the Polish Union, *NSZZ Solidarność* co-released a report with the Heartland Institute, a conservative think tank in the United States of America, that explicitly denied climate science and denounced decarbonisation policies (Thomas and Doerflinger, 2020). At Katowice, Polish trade unions constructed a wall of coal – to highlight the importance of this resource to jobs and livelihoods in the country.

[13] This section was greatly assisted by discussions with Lara Skinner (of The Worker Institute's Labor Leading on Climate Initiative (05/10/2021) and Jeremy Brecher (Labor Network for Sustainability, 24/09/2021).

[14] Lara Skinner of The Worker Institute's Labor Leading on Climate Initiative (05/10/2021).

Chapter 8

[1] I am thankful to Ramón M. Balcázar of *Observatorio Plurinacional de Salares Andinos* for his insights and generosity in explaining the impacts of lithium mining on local communities (20/05/2022).

[2] My thanks to Maria Sanchez-Lopez for her time sharing her extensive insights into the political economies and ecologies of the lithium triangle.

[3] Statistics for both Ethiopia and France are taken from the Global Solar Atlas.

[4] Erik Hagen, Western Sahara Resource Watch (10/09/2021).

[5] I am indebted to a discussion with Erik Hagen of Western Sahara Resource Watch for informing my understanding of the dynamics and complexities of renewable energy projects sited in Western Sahara. I contacted the Moroccan Embassy in London for comment, but I did not receive a response to my emails.

[6] Rebecca Windemer (02/02/2022).

[7] Seb Munoz of War on Want and Andy Whitmore of the London Mining Network (25/11/2021).

[8] Seb Munoz of War on Want and Andy Whitmore of the London Mining Network (25/11/2021).

Chapter 9

[1] This example of Medway Council was found via a tweet by Ed Jennings (2022) on 21 June 2022.

[2] This point is taken from tweets by Alex Chapman (2022) and Simon Evans (2022b) on 26 May 2022.

[3] Truss's government also announced £1.5 billion to improve energy efficiency for around 130,000 properties, prioritising 130,000 low-income households and those living in social housing. Social housing providers and local authorities were to submit bids and compete to secure this funding.

References

Abraham, J. (2017). Just transitions for the miners: Labor environmentalism in the Ruhr and Appalachian coalfields. *New Political Science*, 39(2): 218–240.

Abram, S., Atkins, E., Dietzel, A., Jenkins, K., Kiamba, L., Kirshner, J. et al (2022). Just transition: Pathways to socially inclusive decarbonisation. *Climate Policy*, 22(8): 1033–1049.

Achtenberg, E. (2010). Bolivia: building a state-run lithium industry. *Latin American Bureau*, 16 November. Available at: https://lab.org.uk/bolivia-building-a-state-run-lithium-industry/ (Accessed 27/07/2022).

Adelman, L. (2020). Wind turbine economic impact: Landowner payments. State of Michigan *Clean Energy in Michigan Series*, 1.

Age UK. (2016). Age UK briefing for MPs: Excess winter deaths. *Age UK*, June 2016.

Age UK. (2022). 2.8m older households will still be living in fuel poverty this winter – despite the Government freezing the energy price cap. *Age UK*, 21 September. Available at: https://www.ageuk.org.uk/latest-press/articles/2022/2.8m-older-households-will-still-be-living-in-fuel-poverty-this-winter---despite-the-government-freezing-the-energy-price-cap/ (Accessed 08/12/2022).

Agostinho, A.A., Pelicice, F.M. and Gomes, L.C. (2008). Dams and the fish fauna of the Neotropical region: Impacts and management related to diversity and fisheries. *Braz J Biol*, 68: 1119–1132.

Agren, D. and Stott, M. (2022). Mexico nationalises lithium in populist president's push to extend state control. *Financial Times*, 20 April. Available at: https://www.ft.com/content/5e579b31-c6f0-4911-899a-e2894240ad85 (Accessed 15/12/2022).

Ahlers, R., Budds, J., Joshi, D., Merme, V. and Zwarteveen, M. (2015). Framing hydropower as green energy: Assessing drivers, risks and tensions in the Eastern Himalayas. *Earth Syst. Dyn.*, 6: 195–204.

Aitken, M. (2010). Wind power and community benefits: Challenges and opportunities. *Energy Policy*, 38: 6066–6075.

Ajl, M. (2021). *A People's Green New Deal*. London: Pluto Press.

Allan, J. (2021). Renewable energy is fuelling a forgotten conflict in Africa's last colony. *The Conversation*, 26 November. Available at: https://theconversation.com/renewable-energy-is-fuelling-a-forgotten-conflict-in-africas-last-colony-170995 (Accessed 08/12/2022).

Allan, J., Lemaadel, M. and Lakhal, H. (2022). Oppressive energopolitics in Africa's last colony: Energy, subjectivities and resistance. *Antipode*, 54(1): 44–63.

Allan, J.I. (2020). *The New Climate Activism: NGO Authority and Participation in Climate Change Governances*. Toronto: University of Toronto Press.

Allen, E., Lyons, H. and Stephens, J.C. (2019). Women's leadership in renewable transformation, energy justice and energy democracy: Redistributing power. *Energy Research & Social Science*, 57: 101233.

Allison, J.E., McCrory, K. and Oxnewvad, I. (2019). Closing the renewable energy gender gap in the United States and Canada: The role of women's professional networking. *Energy Research & Social Science*, 55: 35–45.

ALW [Ambition Lawrence Weston]. (2022). *Declaring our Caring: Community Climate Action Plan*. Bristol: ALW.

Aman, M.M., Solangi, K.H., Hossain, M.S., Badarudin, A., Jasmon, G.B., Mokhlis, H. et al (2015). A review of Safety, Health and Environmental (SHE) issues of solar energy system. *Renewable and Sustainable Energy Review*, 31: 1190–1204.

Ambrose, A. (2020a). UK government to subsidise onshore renewable energy projects. *The Guardian*, 24 November. Available at: https://www.theguardian.com/environment/2020/nov/24/uk-government-to-subsidise-onshore-renewable-energy-projects (Accessed 27/07/2022).

Ambrose, A. (2020b). UK government lifts block on new onshore windfarm subsidies. *The Guardian*, 2 March. Available at: https://www.theguardian.com/business/2020/mar/02/uk-government-lifts-block-on-new-onshore-windfarm-subsidies (Accessed 27/07/2022).

Ambrose, A. (2021). Queen's property manager and Treasury to get windfarm windfall of nearly £9bn. *The Guardian*, 8 February. Available at: https://www.theguardian.com/business/2021/feb/08/queens-treasury-windfarm-bp-offshore-seabed-rights (Accessed 27/07/2022).

Amnesty International. (2020). DRC: Alarming research shows long lasting harm from cobalt mine abuses. *Amnesty International*, 6 May. Available at: https://www.amnesty.org/en/latest/news/2020/05/drc-alarming-research-harm-from-cobalt-mine-abuses/ (Accessed 27/07/2022).

Anderson, J.L. (2020). The fall of Evo Morales. *New Yorker*, 23 March. Available at: https://www.newyorker.com/magazine/2020/03/23/the-fall-of-evo-morales (Accessed 08/12/2022).

Anderson, L. (2022). My voters don't care about COP26 ... just their gas bills, says Tory MP. *MailOnline*, 1 January. Available at: https://www.dailym ail.co.uk/debate/article-10361215/My-voters-dont-care-COP26-just-gas-bills-says-Tory-MP-LEE-ANDERSON.html (Accessed 27/07/2022).

Angel, J. (2017). Towards an energy politics in, against, and beyond the state: Berlin's struggles for energy democracy. *Antipode*, 49(3): 557–576.

Angel, J. (2021). New municipalism and the state: Remunicipalising energy in Barcelona, from prosaics to process. *Antipode*, 53(2): 524–545.

Apollo Alliance and Cornell Global Labor Institute. (2009). *Making the Transition: Helping Workers and Communities Retool for the Clean Energy Economy*. Ithaca, NY.: Cornell Global Labor Institute.

Aquatera. (2021). Community owned wind farms have paid their communities 34 times more than commercial counterparts. *Aquatera*, 17 June. Available at: https://www.aquatera.co.uk/news/community-owned-wind-farms-have-paid-their-communities-34-times-more-than-commerc ial-counterparts (Accessed 27/07/2022).

Araújo, C.C. and Wang, J.Y. (2015). The dammed river dolphins of Brazil: impacts and conservation. *Oryx*, 49: 17–24.

Argyll and Bute Council. (nd.). CROP Benefits Community. *Argyll and Bute Council*. Available at: https://www.argyll-bute.gov.uk/crop-benef its-community (Accessed 13/12/2022).

Aronoff, K. (2017). Building power to the people: The unlikely case for utility populism. *Dissent*, Summer 2017. Available at: https://www.diss entmagazine.org/article/the-unlikely-case-for-utility-populism-rural-elect ric-cooperatives (Accessed 27/07/2022).

Aronoff, K., Battistoni, A., Aldana Cohen, D. and Riofrancos, T. (2019). *A Planet to Win: Why We Need a Green New Deal*. London: Verso.

Ash, C., Mackenzie, A., Paxton, L., Smith, M., Stringer, M., Yates, T. et al (2009). Vestas workers and WCA reply to Guardian report, Wednesday 12 August. *Save Vestas*, 16 August. Available at: https://savevestas.wordpr ess.com/2009/08/16/reply-to-guardian-report-wednesday-12-august/ (Accessed 28/07/2022).

Atkins, E. (2020). Contesting the 'greening' of hydropower in the Brazilian Amazon. *Political Geography*, 80: 102179.

Atkins, E. (2022). 'Bigger than Brexit': Exploring right-wing populism and net-zero policies in the United Kingdom. *Energy Research & Social Science*, 90: 102681.

Atkinson-Palombo, C. and Hoen, B. (2014). *Relationship Between Wind Turbines and Residential Property Values in Massachusetts*. A Joint Report of University of Connecticut and Lawrence Berkeley National Laboratory.

Avila, S., Deniau, Y., Sorman, A.H. and McCarthy, J. (2021). (Counter) mapping renewables: Space, justice, and politics of wind and solar power in Mexico. *Environment and Planning E: Nature and Space*, 5(3): 1056–1085.

Avila-Calero, A. (2017). Contesting energy transitions: Wind power and conflicts in the Isthmus of Tehuantepec. *Journal of Political Ecology*, 24: 993.

Bailey, E., Devine-Wright, P. and Batel, S. (2016). Using a narrative approach to understand place attachments and responses to power line proposals: The importance of life-place trajectories. *Journal of Environmental Psychology*, 48: 200–211.

Bailey, I. and Darkal, H. (2018). (Not) talking about justice: Justice self-recognition and the integration of energy and environmental-social justice into renewable energy siting. *Local Environment*, 23(3): 335–351.

Bainton, N., Kemp, D., Lèbre, E., Owen, J.R. and Marston, G. (2021). The energy-extractives nexus and the just transition. *Sustainable Development*, 9: 624–634.

Baker, L. (2021). Procurement, finance and energy transition: Between global processes and territorial realities. *Environment and Planning E: Nature and Space*, 5(4): 1738–1764.

Baker, S. (2019). Anti-resilience: A roadmap for transformational justice within the energy system. *Harvard Civil Rights: Civil Liberties Law Review*, 54: 1–48.

Baker, S. (2021). *Revolutionary Power: An Activists Guide to the Energy Transition*. Washington, DC: Island Press.

Bakker, R.H., Pedersen, E., van der Berg, G.P., Stewart, R.E., Lok, W. and Bouma, J. (2012). Impact of wind turbine sound on annoyance, self-reported sleep disturbance and psychological distress. *Sci. Total Environ*, 425: 42–51.

Bankwatch. (2019). *Western Balkans Hydropower, Who Pays, Who Profits? How Renewables Incentives Have Fed the Small Hydropower Boom and What Needs to Change*. Prague: Bankwatch.

Barca, S. (2015). 'Greening the job: Trade unions, climate change and the political ecology of labour.' In: Bryant, R.L. (ed.) *International Handbook of Political Ecology*. London: Edward Elgar.

Bardazzi, R., Bortolotti, L. and Pazienza, M.G. (2021). To eat and not to heat? Energy poverty and income inequality in Italian regions. *Energy Research & Social Science*, 73: 101946.

Barker, G. (2013a). Gas isn't the bogeyman. *The Guardian*, 12 September. Available at: https://www.theguardian.com/commentisfree/2013/sep/12/gas-cleaner-energy-sector (Accessed 27/07/2022).

Barker, G. (2013b). Speech by Minister of State Greg Barker at the Ground Source Heat Pump Association's fourth technical seminar. *UK Department of Energy and Climate Change*, 5 December. Available at: https://www.gov.uk/government/speeches/ground-source-heat-pump-association (Accessed 08/12/2022).

Barrass, K. (2022). *Local Net Zero Delivery Progress Reports*. Harlow: UK100.

Barrett, J., Pye, S., Betts-Davies, S., Broad, O., Price, J., Eyre, N. et al (2022). Energy demand reduction options for meeting national zero-emission targets in the United Kingdom. *Nature Energy*, 7: 726–735.

Bartholomew, E. (2018). Olympic legacy? 3/4 of the promised 11,000 jobs for 2012 park still don't exist. *Hackney Gazette*, 25 October. Available at: https://www.hackneygazette.co.uk/news/most-of-the-promised-11-000-jobs-for-2012-olympics-3608098 (Accessed 27/07/2022).

Bartik, T.J. (2020). Bringing jobs to people: Improving local economic development policies. Policy Paper no: 2020–023. Kalamazoo, MI: W.E. Upjohn Institute for Employment Research.

Bauwens, T., Schraven, D., Drewing, E., Radtke, J., Holstenkamp, L., Gotchev, B. and Yildiz, O. (2022). Conceptualising community in energy systems: A systematic review of 183 definitions. *Renewable and Sustainable Energy Reviews*, 156: 111999.

BBC. (2013). Italy seizes record assets from wind farm tycoon. *BBC News*, 3 April. Available at: https://www.bbc.co.uk/news/world-europe-22017112 (Accessed 27/07/2022).

BBC. (2017). Kilgallioch wind farm accident death worker identified. *BBC News*, 27 March. Available at: https://www.bbc.co.uk/news/uk-scotland-south-scotland-39404625 (Accessed 27/07/2022).

BBC. (2020). Blyth Power Station to be turned into UK's first 'gigafactory'. *BBC News*, 12 April. Available at: https://www.bbc.co.uk/news/uk-england-tyne-56711116 (Accessed 27/07/2022).

BBC. (2021). Firms fined £900,000 over Ayrshire wind farm worker's death. *BBC News*, 17 November. Available at: https://www.bbc.co.uk/news/uk-scotland-glasgow-west-59324806 (Accessed 08/12/2022).

BBC. (2022a). Ovo Energy sorry over advice to cuddle pets to stay warm. *BBC News*, 11 January. Available at: https://www.bbc.co.uk/news/business-59946622 (Accessed 27/07/2022).

BBC. (2022b). Stars say Suffolk wind farm plans 'anything but green'. *BBC News*, 28 February. Available at: https://www.bbc.co.uk/news/uk-england-suffolk-60560313 (Accessed 27/07/2022).

BCC [Bristol City Council] (2022). Ward profile report: Avonmouth and Lawrence Weston. Bristol City Council, September 2022. Available at: https://www.bristol.gov.uk/files/documents/1960-avonmouth-and-lawrence-weston-ward-profile-report (Accessed 13/01/2022).

Becker, S. (2017). 'Our City, Our Grid: The energy remunicipalisation trend in Germany'. In: Kishimoto, S. and Petitjean, O. *Reclaiming Public Services: How Cities and Citizens are Turning Back Privatisation*. Amsterdam: Transnational Institute.

Becker, S. and Naumann, M. (2017). Energy democracy: Mapping the debate on energy alternatives. *Geography Compass*, 11: e12321.

Bedi, H.P. (2019). "Lead the district into the light": Solar energy infrastructure injustices in Kerala, India. *Global Transitions*, 1: 181–189.

BEIS [Business, Energy and Industrial Strategy, UK Department for]. (2020a). *Digest of UK Energy Statistics (DUKES) 2020*. London: UK Government.

BEIS. (2020b). UK government launches taskforce to support drive for 2 million green jobs by 2030. *BEIS*, 12 November. Available at: https://www.gov.uk/government/news/uk-government-launches-taskforce-to-support-drive-for-2-million-green-jobs-by-2030 (Accessed 28/07/2022).

BEIS. (2021a). *UK Rooftop Solar Behaviour Research: A Report by Basis Social*. BEIS Research Paper, No. 2021/018.

BEIS. (2021b). BEIS Public Attitudes Tracker (March 2021, Wave 37, UK). BEIS, 13 May.

BEIS. (2022a). BEIS Public Attitudes Tracker: Heat and Energy in the Home Spring 2022. BEIS, 16 June.

BEIS. (2022b). Biggest renewables auction accelerates move away from fossil fuels. *BEIS*, 7 July. Available at: https://www.gov.uk/government/news/biggest-renewables-auction-accelerates-move-away-from-fossil-fuels (Accessed 28/07/2022).

BEIS. (2022c). *Annual Fuel Poverty Statistics in England, 2022 (2020 data)*. BEIS, 24 February.

BEIS. (2022d). *Smart Meters in Great Britain, Quarterly Update (March 2022)*. BEIS, 26 May.

BEIS Committee [Business, Energy and Industrial Strategy Committee, UK House of Commons] (2022). Energy pricing and the future of the energy market. Third Report of Session 2022–23. UK House of Commons, 26 July. Available at: https://committees.parliament.uk/work/1698/energy-pricing-and-the-future-of-the-energy-market/publications/ (Accessed 08/12/2022).

Belger, T. (2022). 'Apocalyptic': 500 per cent energy price hikes plunge schools into winter crisis. *Schools Week*, 6 September. Available at: https://schoolsweek.co.uk/school-energy-bills-government-help-apocalyptic-rises/ (Accessed 08/12/2022).

Bell, A. (2022). How realistic is Labour's plan for zero-carbon energy by 2030? *CAPX*, 26 September. Available at: https://capx.co/how-realistic-is-labours-plan-for-zero-carbon-energy-by-2030/ (Accessed 08/12/2022).

Bell, D. and Carrick, J. (2018). 'Procedural environmental justice'. In: Holifield, R., Chakraborty, J., Walker, G. (eds.). *The Routledge Handbook of Environmental Justice*. Abingdon: Routledge.

Bell, K. (2014). *Achieving Environmental Justice: A Cross-National Analysis*. Bristol: Policy Press.

Benchimol, M. and Peres, C.A. (2015) Widespread forest vertebrate extinctions Induced by a mega hydroelectric dam in lowland Amazonia. *PLoS ONE*, 10(7): e0129818.

Bennis, A. (2021). Power surge: How the European Green Deal can succeed in Morocco and Tunisia. *European Council on Foreign Relations*, 26 January. Available at: https://ecfr.eu/publication/power-surge-how-the-european-green-deal-can-succeed-in-morocco-and-tunisia/ (Accessed 17/03/2023).

Bergquist, P., Mildenberger, M. and Stokes, L.C. (2020). Combining climate, economic, and social policy builds public support for climate action in the US. *Environmental Research Letters*, 15(5): 054019.

Bernal, N. (2021). The race to grab All the UK's lithium before it's too late. *Wired*, 5 October. Available at: https://www.wired.co.uk/article/cornwall-lithium (Accessed 28/07/2022).

Bernards, N. (2022). Global capitalism and the scramble for cobalt. *Review of African Political Economy*, 5 January. Available at: https://roape.net/2022/01/05/global-capitalism-and-the-scramble-for-cobalt/ (Accessed 28/07/2022).

Bessi, R. and Navarro, S. (2016). The dark side of clean energy in Mexico. *Avispa Midia*, 14 February. Available at: https://avispa.org/the-dark-side-of-clean-energy-in-mexico/ (Accessed 27/07/2022).

Beynon, H. and Wainwright, H. (1979). *Alternative Planning: the Lucas Combine Committee, A Discussion Paper*.

Beynon, H. and Hudson, R. (2021). *The Shadow of the Mine: Coal and the End of Industrial Britain*. London: Verso.

BHA [British Hydropower Association]. (no date.) *Types of hydro generation*. British Hydropower Association. Available at: https://www.british-hydro.org/types-of-hydro/ (Accessed 27/07/2022).

BHRC [Business and Human Rights Resource Centre]. (2019). Transition Minerals Tracker: Analysis of renewable energy mining companies' human rights practice. *Business and Human Rights Resource Centre*, 5 September. Available at: https://www.business-humanrights.org/en/from-us/briefings/transition-minerals-tracker-analysis-of-renewable-energy-mining-companies-human-rights-practice/ (Accessed 28/07/2022).

Bidwell, D. (2013). The role of values in public beliefs and attitudes towards commercial wind energy. *Energy Policy*, 58: 189–199.

Bidwell, D., Firestone, J. and Ferguson, M.D. (2022). Love thy neighbor (or not): Regionalism and support for the use of offshore wind energy by others. *Energy Research & Social Science*, 90: 102599.

Bird. C.M. and Barnes, J. (2014). Scaling up community activism: the role of intermediaries in collective approaches to community energy. *People Place Policy*, 8(3): 208–221.

Boardman, B. (2009). *Fixing Fuel Poverty: Challenges and Solutions*. London: Routledge.

Bolger, M., Marin, D., Tofighi-Niaki, A. and Seelmann, L. (2021). 'Green mining' is a myth: The case for cutting EU resource consumption. Brussels: European Environmental Bureau and Friends of the Earth Europe.

Bolton, P. and Stewart, I. (2023). Domestic energy prices. London: Domestic energy prices. Available at: https://commonslibrary.parliament.uk/research-briefings/cbp-9491/ (Accessed 18/03/2023).

Bomberg, E. and McEwen, N. (2012). Mobilizing community energy. *Energy Policy*, 51: 435–444.

Bonneuil, C., Choquet, P-L. and Franta, B. (2021). Early warnings and emerging accountability: Total's responses to global warming, 1971–2021. *Global Environmental Change*, 71: 102386.

Boon, F.P. and Dierperink, C. (2014). Local civils society based renewable energy organisations in the Netherlands: Exploring the factors that stimulate their emergence and development. *Energy Policy*, 69: 297–307.

Boren, Z. (2021). Leading housebuilder pushed for weaker climate targets. *Unearthed*, 5 July. Available at: https://unearthed.greenpeace.org/2021/07/05/housing-net-zero-climate-target-lobbying/ (Accessed 13/12/2022).

Bouzarovski, S. (2022). Just Transitions: A political ecology critique. *Antipode*, 54(4): 1003–1020.

Bouzarovski, S. and Simcock, N. (2017). Spatializing energy justice, *Energy Policy*, 107: 640–648.

Braunholtz-Speight, T., Mander, S., Hannon, M., Hardy, J., McLachlan, C., Manderson, E. and Sharmina, M. (2018). The evolution of community energy in the UK. UK Energy Research Centre. Available at: https://ukerc.ac.uk/publications/evolution-of-community-energy-in-the-uk/ (Accessed 08/12/2022).

BRE [BRE National Solar Centre]. (2016). *Solar PV on Commercial Buildings: A guide for Owners and Developers*. BRE National Solar Centre.

Breeze, P. (2019). *Power Generation Technologies*. Amsterdam: Elsevier.

Brennan, N. and van Rensburg, T.M. (2016). Wind farm externalities and public preferences for community consultation in Ireland: A discrete choice experiments approach. *Energy Policy*, 94: 355–365.

Bridge, G., Bouzarovski, S., Bradshaw, M. and Eyre, N. (2013). Geographies of energy transition: Space, place and the low-carbon economy. *Energy Policy*, 53: 331–340.

Bridge, G., Barr, S., Bouzarovski, S., Bradshaw, M., Brown, E., Bulkeley, H. and Walker, G. (2018). *Energy and Society: A Critical Perspective*. London: Routledge.

Bridge, G. and Gailing, L. (2020). New energy spaces: Towards a geographical political economy of energy transition. *Environment and Planning A: Economy and Space*, 52(6): 1037–1050.

Brisbois, M.C. (2019). Powershifts: A framework for assessing the growing impact of decentralized ownership of energy transitions on political decision-making. *Energy Research & Social Science*, 50: 151–161.

Bristol Energy. (2019). Bristol Energy Annual Results 2018/19. *Bristol Energy*, 8 August. Available at: https://www.bristol-energy.co.uk/bristol-energy-annual-results-201819 (Accessed 28/07/2022).

Bristol Energy Cooperative. (2020). A busy year for the Bristol Community Hydro Scheme. *Bristol Energy Cooperative*, 16 October. Available at: https://bristolenergy.coop/a-busy-year-for-the-bristol-community-hydro-scheme/ (Accessed 28/07/2022)

Bristol Energy Cooperative. (2022). Going big at Bottle Yard Studios. *Bristol Energy* Cooperative, 23 June. Available at: https://bristolenergy.coop/going-big-at-bottle-yard-studios/?mc_cid=1f46f57204&mc_eid=c13066c861 (Accessed 28/07/2022)

Brock, A., Sovacool, B.K. and Hook, A. (2021) Volatile photovoltaics: Green industrialization, Sacrifice zones, and the political ecology of solar energy in Germany, *Annals of the American Association of Geographers*, 111(6): 1756–1778.

Brock, G. (2009). *Global Justice: A Cosmopolitan Account*. Oxford: Oxford University Press.

Brooks, E. and Davoudi, S. (2014). Climate justice and retrofitting for energy efficiency: Examples from the UK and China. *disP - The Planning Review*, 50(3): 101–110.

Brown, D., Hall, S. and Davis, M.E. (2020a). What is prosumerism for? Exploring the normative dimensions of decentralised energy transitions. *Energy Research & Social Science*, 66: 101475.

Brown, D. and Bailey, T. (2022). Cheaper Bills, Warmer Homes. Available at: https://www.cheaperbillswarmerhomes.org/ (Accessed 08/12/2022).

Brown, M.A., Soni, A., Lapsa, M.V., Southworth, K. and Cox, M. (2020b). High energy burden and low-income energy affordability: conclusions from a literature review. *Prog. Energy*, 2(4): 042003.

Brulle, R.J. (2018). The climate lobby: a sectoral analysis of lobbying spending on climate change in the US, 2000 to 2016. *Climatic Change*, 149: 289–303.

Buchan, L. (2021). Keir Starmer says Labour will not nationalise 'Big Six' energy firms. *The Mirror*, 26 September. Available at: https://www.mirror.co.uk/news/politics/keir-starmer-says-labour-not-25074686 (Accessed 28/07/2022).

Buli, N. and Jacobsen, S. (2021). Analysis: Weak winds worsened Europe's power crunch; utilities need better storage. *Reuters*, 22 December. Available at: (Accessed 18/03/2023).

Bullard, R.D. (1993). *Confronting Environmental Racism: Voices from the Grassroots*. Boston: South End Press.

Bump, P. (2013). Somehow, the renewable sector in Sicily was infiltrated by the mob. *Grist*, 23 January. Available at: https://grist.org/business-tec hnology/somehow-the-renewable-sector-in-sicily-was-infiltrated-by-the-mob/ (Accessed 28/07/2022).

Burkett, E. (2003). A mighty wind. *The New York Times Magazine*, 15 June. Available at: https://www.nytimes.com/2003/06/15/magazine/a-mig hty-wind.html (Accessed 28/07/2022).

Burns, R. (2021). The subprime solar trap for low-income homeowners. *Bloomberg*, 6 April. Available at: https://www.bloomberg.com/news/featu res/2021-04-06/the-subprime-solar-trap-for-low-income-homeowners (Accessed 28/07/2022).

C3E International. (2019). *Status Report on Gender Equality in the Energy Sector*. C3E International.

Campbell, P., Dempsey, H. and Agnew, H. (2022a). The EV battery race: inside the struggles of Britishvolt. *Financial Times*, 28 September. Available at: https://www.ft.com/content/7cd57531-1c54-4d32-955c-d185dcea0621 (Accessed 08/12/202).

Campbell, P., Dempsey, H. and Agnew, H. (2022b). Britishvolt on brink after government rejects rescue plea. *Financial Times*, 31 October. Available at: https://www.ft.com/content/adf3ee60-4734-4e0c-b27b-55842136d 3f7 (Accessed 08/12/2022).

Cannon, M. and Thorpe, J. (2020). Preston model: Community wealth generation and a local cooperative economy. Participatory Economic Alternatives Case Summary, 20.

Capellán-Pérez, I., Campos-Celador, Á. and Terés-Zubiaga, J. (2018). Renewable energy cooperatives as an instrument towards the energy transition in Spain. *Energy Policy*, 123: 215–229.

Cardwell, D. (2014). U.S. imposes steep tariffs on Chinese solar panels. *New York Times*, 16 December. Available at: https://www.nytimes.com/ 2014/12/17/business/energy-environment/-us-imposes-steep-tariffs-on-chinese-solar-panels.html (Accessed 08/12/2022).

Carlson, J.D. (2019). Renewable energy and class struggles: Slurry and stratification in Germany's energy transition. *Rachel Carson Center Perspectives: Transformations in Environment and Society*: 2019/2: 47–56.

Carter, L. (2016). British Gas owner Centrica funding climate denial group linked to Trump. *Unearthed*, 15 December. Available at: https://unearthed. greenpeace.org/2016/12/15/british-gas-centrica-donates-climate-denial-linked-trump/ (Accessed 08/12/2022).

Cashmore, M., Rudolph, D., Larsen, S.V. and Nielsen, H. (2019). International experiences with opposition to wind energy siting decisions: Lessons for environmental and social appraisal. *Journal of Environmental planning and Management*, 62(7): 1109–1132.

Castán Broto, V. and Calvet, M.S. (2020). Sacrifice zones and the construction of urban energy landscapes in Concepción, Chile. *Journal of Political Ecology*, 27: 279–299.

Cederlöf, G. (2020). Maintaining power: Decarbonisation and recentralisation in Cuba's Energy Revolution. *Trans Inst Br Geogr.*, 45: 81–94.

Cha, J.M. (2020). A just transition for whom? Politics, contestation and social identity in the disruption of coal in the Power River Basin. *Energy Research & Social Science*, 69: 101657.

Chaffin, J. (2012). Chinese solar groups in EU trade spat. *Financial Times*, 24 July. Available at: https://www.ft.com/content/2289a744-d5be-11e1-a5f3-00144feabdc0 (Accessed 08/12/2022).

Chapman, A. (2022). Twitter, 26 May. Available at: https://twitter.com/chappersmk/status/1529838700795936769 (Accessed 28/07/2022).

Charmorel, P. (2019). Macron vs the Yellow Vests. *Journal of Democracy*, 30(4): 48–62.

Chen, H., Ohura, J., Maruyama, M. (2022). Japan tells millions to save electricity as record heat wave strains power supply. *CNN*, 28 June. Available at: https://edition.cnn.com/2022/06/28/asia/japan-heatwave-air-conditioning-power-electricity-shortage-climate-change-intl-hnk/index.html (Accessed 28/07/2022).

CIEL [Center for International Environmental Law]. (2016.) Barro Blanco hydroelectric dam threatens Indigenous communities, Panama. *Center for International Environmental Law*. Available at: https://www.ciel.org/project-update/barro-blanco/ (Accessed 28/07/2022).

Ciplet, D. and Harrison, J.L. (2020). Transition tensions: Mapping conflicts in movements for a just and sustainable transition. *Environmental Politics*, 29(3): 1–22.

Citizens Advice. (2021). Catalogue of errors at Ofgem leaves consumers with multi-billion pound bill. *Citizens Advice*, 9 December. Available at: https://www.citizensadvice.org.uk/about-us/about-us1/media/press-releases/catalogue-of-errors-at-ofgem-leaves-consumers-with-multi-billion-pound-bill/ (Accessed 08/12/2022).

Citizens Advice. (2022). *Out of the Cold? Helping People on Prepayment Meters Stay Connected this Winter*. London: Citizens Advice.

Citizens Advice. (2023). Millions left in the cold and dark as someone on a prepayment meter cut off every 10 seconds, reveals Citizens Advice. *Citizens Advice*, 12 January. Available at: https://www.citizensadvice.org.uk/about-us/about-us1/media/press-releases/millions-left-in-the-cold-and-dark-as-someone-on-a-prepayment-meter-cut-off-every-10-seconds-reveals-citizens-advice/ (Accessed 13/01/2023).

Clancy, J. and Feenstra, M. (2019). Women, gender equality and the energy transition in the EU. European Union Policy Department for Citizens' Rights and Constitutional Affairs, May 2019.

Clark, H. (2021). Examining the end of the furlough scheme. *House of Commons Library*, 15 November. Available at: https://commonslibrary.parliament.uk/examining-the-end-of-the-furlough-scheme/ (Accessed 09/12/2022).

Clarke, E. (2019). Question Time viewers slam Emma Barnett after host asked Angela Rayner if Labour would 'nationalise sausages'. *Evening Standard*, 10 December. Available: https://www.standard.co.uk/news/politics/question-time-host-emma-barnett-labour-nationalise-sausages-a4309576.html (Accessed 28/07/2022).

CLES [Centre for Local Economic Strategies]. (2020). *Own the Future: A Guide for New Local Economies*. Manchester: Centre for Local Economic Strategies.

Cleve Hill Solar. (n.d.) About us. Available at: https://www.clevehillsolar.com/ (Accessed 26/07/2022).

Cohen, J.J., Azarova, V., Kollman, A. and Reichl, J. (2021). Preferences for community renewable energy investments in Europe. *Energy Economics*, 1000: 105386.

Community Energy England. (2022). *Community Energy State of the Sector, 2022 Report*. Community Energy England.

Connecticut House Democrats. (2021). Press release: Rights for renewable energy workers. Available at: https://patch.com/connecticut/hamden/rights-renewable-energy-workers-passes (Accessed 27/07/2022).

Connelly, T. (2021). Samsø: Generating energy, and possibilities. *RTE*, 4 November. Available at: https://www.rte.ie/news/analysis-and-comment/2021/1103/1257733-samso-denmark-climate/ (Accessed 28/07/2022).

Conway, E.M. and Oreskes, N. (2012). *Merchants of Doubt: How a Handful of Scientists Obscured the Truth on Issues from Tobacco Smoke to Global Warming*. London: Bloomsbury.

Cooke, F.M., Nordensvard, J., Saat, G.B., Urban, F. and Siciliano, G. (2017). The limits of social protection: The case of hydropower dams and Indigenous peoples' land. *Asia & the Pacific Policy Studies*, 4(3): 437–450.

Coolsaet, B. and Néron, P-Y. (2021). 'Recognition and environmental justice. In Coolsaet, B. (ed.) *Environmental Justice: Key Issues*. Abingdon: Earthscan from Routledge.

Copena, D. and Simon, X. (2018). Wind farms and payments to landowners: Opportunities for rural development for the case of Galicia. *Renewable and Sustainable Energy Reviews*, 95: 38–47.

Coward, R. (2018). Guest blog: Graveney Marshes by Rosalind Coward. *Markavery.info*, 16 October. Available at: https://markavery.info/2018/10/16/guest-blog-graveney-marshes-by-rosalind-coward/ (Accessed 25/07/2022).

Cowell, R., Bristow, G. and Munday, M. (2012). *Wind energy and justice for disadvantage communities*. Joseph Rowntree Foundation. Available at: https://www.jrf.org.uk/sites/default/files/jrf/migrated/files/wind-farms-communities-summary.pdf (Accessed 26/07/2022).

Coy, D., Malekpour, S., Saeri, A.K. and Dargaville, R. (2021). Rethinking community empowerment in the energy transformation: A critical review of the definitions, drivers and outcomes. *Energy Research & Social Science*, 72: 101871.

Craig, J. (2022). Liz Truss to ban solar projects on farms as Tory MP warns plan is 'unwise'. *Sky News*, 11 October. Available at: https://news.sky.com/story/liz-truss-to-ban-solar-projects-on-farms-as-tory-mp-warns-plan-is-unwise-12717624 (Accessed 09/12/2022).

Cribb, J., Hood, A. and Hoyle, J. (2018). The decline of homeownership among young adults. London: Institute for Fiscal Studies.

Crowe, J.A. and Li, R. (2020). Is the just transition socially accepted? Energy, history, place and support for coal and solar in Illinois, Texas and Vermont. *Energy Research & Social Science*, 59: 101309.

Crown Estate, The. (2021). Offshore Wind Leasing Round 4 – Tender process outcome. Available at: https://www.thecrownestate.co.uk/media/3920/round-4-tender-outcome-dashboard.pdf.

Crown Estate Scotland. (2022). ScotWind offshore wind leasing delivers major boost to Scotland's net zero aspirations. *Crown Estate Scotland*, 17 January. Available at: https://www.crownestatescotland.com/news/scotwind-offshore-wind-leasing-delivers-major-boost-to-scotlands-net-zero-aspirations (Accessed 28/07/2022).

Cuesta-Fernández, I., Belda-Miquel, S. and Tormo, C.C. (2020). Challengers in enery transitions beyond renewable energy cooperatives: Community-owned electricity distribution cooperatives in Spain. Innovation: *The European Journal of Social Science Research*, 33(2): 140–159.

Cuff, M. (2022). Seven in 10 Britons believe expansion of UK's renewable energy can solve cost of living crisis. *iNews*, 19 July. Available at: https://inews.co.uk/news/renewable-energy-cost-living-crisis-1748805 (Accessed 28/07/2022).

Cullinane, J. (2021). Scotland's red council. *Tribune*, 7 February. Available at: https://tribunemag.co.uk/2021/02/scotlands-red-council (Accessed 17/03/2023).

Cumbers, A. (2016). Economic democracy: Reclaiming public ownership as the pragmatic left alternative. *Juncture*, 22(3): 324–328.

Cumbers, A. and Paul, F. (2022). Remunicipalisation, mutating neoliberalism, and the conjuncture. *Antipode*, 54(1): 197–217.

D'Ippolito, M. (2021). Ondata di incendi in Sicilia, la pista del fotovoltaico. *EPOCH Times*, 1 August. Available at: https://www.epochtimes.it/news/ondata-incendi-sicilia-pista-fotovoltaico (Accessed 28/07/2022).

Daisley, S. (2022). Does Suella Braverman understand welfare? *The Spectator*, 12 July. Available at: https://www.spectator.co.uk/article/does-suella-braverman-understand-welfare/ (Accessed 13/01/2023).

Daly, P. (2019). The reason why Grimsby offshore wind sector is allowed to pay staff less than the minimum wage. *Grimsby Live*, 2 May. Available at: https://www.grimsbytelegraph.co.uk/news/grimsby-news/offshore-wind-minimum-wage-grimsby-2819861 (Accessed 28/07/2022).

Davies, R. (2022). National Grid to be partly nationalised to help reach net zero targets. *The Guardian*, 6 April. Available at: https://www.theguardian.com/business/2022/apr/06/national-grid-to-be-partially-nationalised-to-help-reach-net-zero-targets (Accessed 28/07/2022).

Davis, M. (2021). Community municipal investments: Accelerating the potential of local net zero strategies. Leeds: University of Leeds.

Dawley, S. (2014). Creating new paths? Offshore wind, policy activism, and peripheral region development. *Economic Geography*, 90(1): 91–112.

Dawson, A. (2020). *People's Power: Reclaiming the Energy Commons*. New York: OR Books.

Dayaneni, G. (2009). Carbon fundamentalism vs. climate justice. *Race, Poverty and the Environment*, 16(2): 7–11.

Dear, M. (1992). Understanding and overcoming the NIMBY syndrome. *Journal of the American Planning Association*, 58: 288–300.

DECC Ireland [Environment, Climate and Communication, Irish Government Department of the] (2022). National Retrofit Plan. Dublin: Government of Ireland.

DECC UK [Energy and Climate Change, UK Department for]. (2014). *Community Energy Strategy: Full Report*. London: UK Government.

Deignan, B., Harvey, E. and Hoffman-Goetz, L. (2013). Fright factors about wind turbines and health in Ontario newspapers before and after the Green Energy Act. *Health, Risk & Society*, 15(3): 234–250.

Delanty, G. (2014). The prospects of cosmopolitanism and the possibility of global justice. *Journal of Sociology*, 50(2): 213–228.

Delina, L.L. (2018). Whose and what futures? Navigating the contested co-production of Thailand's energy socio-technical imaginaries. *Energy Research & Social Science*, 36: 48–56.

Dempsey, H. and Ruehl, M. (2022). Indonesia considers Opec-style cartel for battery metals. *Financial Times*, 31 October. Available at: https://www.ft.com/content/0990f663-19ae-4744-828f-1bd659697468 (Accessed 09/12/2022)

Deppisch, L. (2021). "Where people in the countryside feel left behind populism has a clear path" – an analysis of the popular media discourse on how infrastructure decay, fear of social decline, and right-wing (extremist) values contribute to support for right-wing populism. Thünen Institute of Rural Studies, Braunschweig: Thünen Working Paper, 119a.

Devine-Wright, P. (2009). Rethinking NIMBYism: The role of place attachment and place identity in explaining place protective action. *Journal of Community and Applied Social Psychology* 19(6): 426–441.

Devine-Wright, P. (2011). Place attachment and public acceptance of renewable energy: A tidal energy case study. *Journal of Environmental Psychology*, 331: 336–342.

Devine-Wright, P. (2013). Think global, act local? The relevance of place attachments and place identifies in a climate changed world. *Global Environmental Change*, 23: 61–69.

Devine-Wright, P. (2015). Local attachments and identities: A theoretical and empirical project across disciplinary boundaries. *Progress in Human Geography*, 39(4): 527–530.

Devine-Wright, P. (2019). Community versus local energy in a context of climate emergency. *Nature Energy*, 4: 894–896.

Devine-Wright, P. and Howes, Y. (2010). Disruption to place attachment and the protection of restorative environments: A wind energy case study. Journal of Environmental Psychology, 30: 271–280.

Diesendorf, M. (2022). Scenarios for mitigating CO_2 emissions from energy supply in the absence of CO_2 removal. *Climate Policy*, 22(7): 882–892.

Dietzel, A. (2022). Non-state climate change action: Hope for just response to climate change? *Environment Science and Policy*, 131: 128–134.

Dinneen, J. (2022). Drought, not lithium mining, is drying out Chile's largest salt flat. *New Scientist*, 4 November. Available at: https://www.newscient ist.com/article/2345815-drought-not-lithium-mining-is-drying-out-chi les-largest-salt-flat/ (Accessed 09/12/2022).

Diski, R. (2021a). Preparing for a just transition in Yorkshire and the Humber. *New Economics Foundation*, 10 June. Available at: https://newec onomics.org/2021/06/preparing-for-a-just-transition-in-yorkshire-and- the-humber (Accessed 28/07/2022).

Diski, R. (2021b). Earth, wind and fire. *The New Economics Zine*, 4.

Dixon, R. (2021). Just Transition for people and the climate. *Friends of the Earth Scotland*, 13 October. Available at: https://foe.scot/just-transition- for-people-and-the-climate/ (Accessed: 25/07/2022).

Dobson, P. (2022). Offshore wind farms paid 'measly' sum to communities. *The Ferret*, 29 May. Available at: https://theferret.scot/offshore-wind-mea sly-sum-communities-last-year/ (Accessed 28/07/2022).

Dobson, P. and Matijevic, P. (2022). Wind farms to pay 'loose change' to Scots while fuel bills soar. *The Ferret*, 10 March. Available at: https://thefer ret.scot/wind-farms-pay-loose-change-scots-fuel-bills-soar/ (Accessed 28/07/2022).

Dogmus, Ö. and Nielsen, J.Ø. (2019). Is the hydropower boom actually taking place? A case study of a south east European country, Bosnia and Herzegovina. *Renewable and Sustainable Energy Reviews*, 110: 278–289.

Dogmus, Ö. and Nielsen, J.Ø. (2020). The on-paper hydropower boom: A case study of corruption in the hydropower sector in Bosnia and Herzegovina. *Ecological Economics*, 172: 106630.

Driscoll, J. (2021). Focusing on local wealth creation to level up the North. The Royal Society of Arts, 25 October. Available at: https://www.thersa. org/reports/regional-wealth-generation (Accessed 09/12/2022).

Droubi, S., Heffron, R.J. and McCauley, D. (2022). A critical review of energy democracy: A failure to deliver justice?. *Energy Research & Social Science*, 86: 102444.

Dudley, M. (2020). The limits of power: Wind energy, Orkney, and the post-war British state. *Twentieth Century British History*, 31(3): 316–339.

Dudley, M. (2021). When's a gale a gale? Understanding wind as an energetic force in mid-twentieth century Britain. *Environmental History*, 26: 671–695.

Duff, R. (2022). £454,000 invested into Orkney's hydrogen infrastructure. *Energy Voice*, 18 August. Available at: https://www.energyvoice.com/ren ewables-energy-transition/436417/454000-invested-into-orkneys-emec/ (Accessed 09/12/2022).

Dunlap, A. (2021a). Spreading 'green' infrastructural harm: Mapping conflicts and socio-ecological disruptions within the European Union's transnational energy grid. *Globalizations*, OnlineFirst.

Dunlap, A. (2021b). 'Does renewable energy exist? Fossil Fuel+ technologies and the search for renewable energy'. In: S. Batel, D. Rudolph (eds.). *A Critical Approach to the Social Acceptance of Renewable Energy Infrastructures*. London: Palgrave Macmillan.

Dunlap, A. and Arce, M.C. (2021). 'Murderous energy' in Oaxaca, Mexico: Wind factories, territorial struggle and social warfare. *The Journal of Peasant Studies*, 49(2): 455–480.

Dunlap, A. and Laratte, L. (2022). European Green Deal necropolitics: Exploring 'green' energy transition, degrowth and infrastructural colonization. *Political Geography*, 97: 102640.

Dunlap, A. and Marin, D. (2022). Comparing coal and 'transition materials'? Overlooking complexity, flattening reality and ignoring capitalism. Energy Research & Social Science, 89: 102531.

Dwyer, J. and Bidwell, D. (2019). Chains of trust: Energy justice, public engagement, and the first offshore wind farm in the United States. *Energy Research & Social Science*, 47: 166–176.

ECCHR [European Center for Constitutioanl and Human Rights]. (2021). Press release: Human rights violations off the rack: Dutch and US brands allegedly rely on forced labor. *European Center for Constitutioanl and Human Rights*, 2 December. Available at: https://www.ecchr.eu/en/press-release/ human-rights-violations-off-the-rack/ (Accessed 17/03/2023).

ECIU [Energy and Climate Intelligence Unit]. (2022). *Levelling Up or Letting Down? Tackling poor quality homes in marginal constituencies could swing election success.* London: Energy and Climate Intelligence Unit.

Econie, A. and Doughtery, M.L. (2019). Contingent work in the US recycling industry: Permatemps and precarious green jobs. *Geoforum*, 99: 132–141.

Ediger, V. and Bowlus, J. (2019). A farewell to King Coal: Geopolitics, Energy Security, and the Transition to Oil, 1898-1971. *The Historical Journal*, 62(2), 427–449.

Edwards, T. (2022). UK Government must spend £20bn on energy battery storage to meet 2030 renewables targets. *Cornwall Insight*, 12 May. Available at: https://www.cornwall-insight.com/press/uk-government-must-spend-20bn-on-energy-battery-storage-to-meet-2030-renewables-targets/ (Accessed 28/07/2022).

EEIG [Energy Efficiency Infrastructure Group]. (2022). *The Energy Efficiency Investment Imperative.* Energy Efficiency Infrastructure Group.

Effern, H. (2019). Stadtwerke kaufen Solarparks in ganz Deutschland. *Süddeutsche Zeitung*, 29 September. Available at: https://www.sueddeutsche.de/muenchen/stadtwerke-muenchen-solarparks-1.4620372 (Accessed 28/07/2022).

Ek, K. and Persson, L. (2014). Wind farms – where and how to place them? A choice experiment approach to measure consumer preferences for characteristics of wind farm establishments in Sweden. *Ecol. Econ.*, 105: 193–203.

Ellis, G. and Ferraro, G. (2016). *The Social Acceptance of Wind Energy: Where We Stand and the Path Ahead.* European Commission, JRC Science for Policy Report.

Elson, A. (2022). Energy bills: We won't pay, insist 1.7m planning to stop direct debits. *The Times*, 2 September. Available at: https://www.thetimes.co.uk/article/energy-bills-we-won-t-pay-insist-1-7m-planning-to-stop-direct-debits-pss85pwk7 (Accessed 18/03/2023).

Emden, J. and Rankin, L. (2022). *Pump up the Volume: A Comprehensive Plan to Decarbonise the UK's Homes.* London: IPPR.

Emelianoff,, C. and Wernert, C. (2019). Local energy, a political resource: Dependencies and insubordination of an urban "Stadtwerk" in France (Metz, Lorraine). *Local Environment*, 24(1): 1035–1052.

End Fuel Poverty Coalition. (2022). About fuel poverty. *End Fuel Poverty Coalition.* Available at: https://www.endfuelpoverty.org.uk/about-fuel-poverty/ (Accessed 09/12/2022).

Enercoop. (nd). Notre soutien aux projets : Appartant aux acteurs locaux. *Enercoop.* Available at: https://www.enercoop.fr/la-production-de-notre-electricite/electricite-par-et-pour-les-citoyens (Accessed 28/07/2022).

Enercoop. (2021). 100,000 clients Enercoop : Merci pour votre soutien. *Enercoop*, 17 February. Available at: https://www.enercoop.fr/blog/act ualites/nationale/100-000-clients-enercoop-merci-pour-votre-soutien (Accessed 28/07/2022).

Eneveoldsen, P., Permien, F-H., Bakhtaoui, I., von Krauland, A-K., Jacobsen, M.Z., Xydis, G., Sovacool, B.K., Valentine, S.V., Luecht, D. and Oxley, G. (2019). How much wind power potential does Europe have? Examining European wind power potential with an enhanced socio-technical atlas. *Energy Policy*, 132: 1092–1100.

Environmental Audit Committee, UK. (2021). Technological Innovations and Climate Change inquiry: Removing the barriers to the development of community energy. London: Environmental Audit Committee. Available at: https://committees.parliament.uk/publications/5718/documents/ 56323/default/ (Accessed 09/12/2022).

Erlewein, A. (2018). 'The promotion of dams through the Clean Development Mechanism: Between sustainable climate protection and carbon colonialism'. In Nüsser, M. (ed). *Large Dams in Asia: Contested Environments between Technological Hydroscapes and Social Resistance.* Berlin: Springer.

Erneuerbare Energien. (2016). Wie viel Wertschöpfung bringt ein regionaler Windpark? *Erneuerbare Energien*, 8 June. Available at: https://www.erne uerbareenergien.de/energiemaerkte-weltweit/akzeptanz-und-buerger beteiligung-wie-viel-wertschoepfung-bringt-ein (Accessed 28/07/2022).

ESI Africa. (2021). 92,000 jobs potentially at stake because of the energy transition. *ESI Africa*, 26 November. Available at: https://www.esi-africa. com/features-analysis/92-000-jobs-potentially-at-stake-because-of-the-energy-transition/ (Accessed 17/03/2023).

ESMAP. (2020). *Global Photovoltaic Power Potential by Country*. Washington, DC.: World Bank.

ETIP Wind. (2019). *How Wind is Going Circular: Blade Recycling.* Brussels: ETIP Wind.

ETUC [European Trade Union Confederation]. (2018). Spain guarantees a just transition for miners. *European Trade Union Confederation*. Available at: https://www.etuc.org/en/spain-guarantees-just-transition-miners (Accessed 28/07/2022).

European Climate Foundation. (2021). Europeans support new wind and solar projects in their local area. *European Climate Foundation*, 20 October. Available at: https://europeanclimate.org/resources/europeans-support-new-wind-and-solar-projects-in-their-local-area/ (Accessed 25/07/2022).

Evans, G. (2007). A just transition from coal to renewable energy in the Hunter Valley of New South Wales. Australia. *Int. J. Environ. Workplace Employ.*, 3: 175–194.

Evans, L.J. (2020). Adaptation, Governance and Industrial Diversification: North Sea Ports and the Growth of Offshore Wind. PhD Thesis submitted to Centre for Urban and Regional Development Studies (CURDS), School of Geography, Politics and Sociology, Newcastle University.

Evans, S. (2022a). Analysis: Cutting the 'green crap' has added £2.5bn to UK energy bills. *Carbon Brief*, 20 January. Available at: https://www.carbonbrief.org/analysis-cutting-the-green-crap-has-added-2-5bn-to-uk-energy-bills/ (Accessed 28/07/2022).

Evans, S. (2022b). Twitter, 26 May. Available at: https://twitter.com/DrSimEvans/status/1529829654068084739 (Accessed 28/07/2022).

Evans, G. and Phelan, L. (2016). Transition to a post-carbon society: Linking environmental justice and just transition discourse. *Energy Policy*, 99: 329–339.

Everard, M. (2013). *The Hydropolitics of Dams: Engineering or Ecosystems?* London: Zed Books.

Eyre, N., Sorrell, S., Guertler, P. and Rosenow, J. (2017). Unlocking Britain's First Fuel: The potential for energy savings in UK housing. London: *UK Energy Research Centre*.

Falxa-Raymond, N., Svendsen, E. and Campbell, L.K. (2013). From job training to green jobs: a case study for a young adult employment program centered on environmental restoration in New York City, US. *Urban Forestry & Urban Greening*. 12: 287–295.

Farrell, C. (2012) A just transition: Lessons learned from the environmental justice movement. *Duke Forum for Law & Social Change*, 4(45): 45–63.

Farthing, L. and Fabricant, N. (2018). *Open Veins* revisited: Charting the social, economic, and political contours of the new extractivism in Latin America. *Latin American Perspectives*, 45(5): 4–17.

Fearnside, P.M. (1995) Hydroelectric dams in the Brazilian Amazon as sources of 'greenhouse' gases. *Environ Conserv*, 22: 7–19.

Fearnside, P.M. (2016). Greenhouse gas emissions from Brazil's Amazonian hydroelectric dams. *Environmental Research Letters*, 11(1): 011002.

Ferri, P. (2020). Todas las violencias de México en la matanza de San Mateo del Mar. *El País*, 26 June. Available at: https://elpais.com/internacional/2020-06-25/todas-las-violencias-de-mexico-en-la-matanza-de-san-mateo-del-mar.html (Accessed 28/07/2022).

Ferris, N. (2022). Weekly data: Renewables industry now employs 0.7% of the workforce in EU and China. *Energy Monitor*, 17 October. Available at: https://www.energymonitor.ai/tech/renewables/weekly-data-renewables-industry-now-employs-0-7-of-the-workforce-in-eu-and-china

Ferroukhi, R., Garcia Casals, X. And Parajuli, B. (2020). *Measuring the Socio-economics of Transition: Focus on Jobs*. Abu Dhabi: IRENA. Available at: https://www.irena.org/-/media/Files/IRENA/Agency/Publication/2020/Feb/IRENA_Transition_jobs_2020.pdf (Accessed 27/07/2022).

Finley-Brook, M. and Holloman, E.L. (2016). Empowering energy justice. *Int J Environ Res Public Health*, 13(9): 926.

Firestone, J. and Kempton, W. (2007). Public opinion about large offshore wind power: Underlying factors. *Energy Policy*, 35: 1584–1598.

Firestone, J., Kempton, W. and Krueger, A. (2009). Public acceptance of offshore wind power projects in the US. *Wind Energy*, 12: 183–202.

Firestone, J., Bates, A. and Knapp, L.A. (2015). See me, feel me, touch me, heal me: Wind turbines, culture, landscapes and sound impressions. *Land Use Policy*, 46: 241–249.

Fletcher, R., Dressler, W.H., Anderson, Z.R. and Büscher, B. (2019). Natural capital must be defended: green growth as neoliberal biopolitics. *The Journal of Peasant Studies*, 46(5): 1068–1095.

Flin, B. (2021). 'Like putting a lithium mine on Arlington cemetery': the fight to save sacred land in Nevada. *The Guardian*, 2 December. Available at: https://www.theguardian.com/us-news/2021/dec/02/thacker-pass-lith ium-mine-fight-save-sacred-land-nevada (Accessed 28/07/2022).

FOE Scotland [Friends of the Earth Scotland]. (2020). Just transition partnership: 2021 manifesto. *Friends of the Earth Scotland*, September. Available at: https://foe.scot/resource/just-transition-partnership-manife sto/ (Accessed: 25/07/2022).

Ford, R. (2022). *Words and Waves: Ecological Dialogism as an Approach to Discourse, Community, and Marine Renewable Energy in Orkney*. PhD Thesis submitted to University of Highlands and Islands.

Fornahl, D., Hassink, R., Klaerding, C., Mossig, I., and Schröder, H. (2012). From the old path of shipbuilding onto the new path of offshore wind energy? The case of North Germany. *European Planning Studies*, 20(5): 835–855.

Foster, S. (1998). Justice from the ground up: Distributive inequities, grassroots resistance and the transformative politics of the environmental justice movement, *California Law Review*, 86: 775.

Fox, N. (2018). 'Knowing energy': The experience of prosumers living on a social housing estate'. In: Lloyd, H. (ed.). A Distributed Energy Future for the UK: An Essay Collection. London: Institute for Public Policy Research.

Franquesa, J. (2018). *Power Struggles: Dignity, Value and the Renewable Energy Frontier in Spain*. Bloomington, ID.: Indiana University Press.

Fraser, D. (2021). How do we make homes fit for net zeroes? *BBC News*, 6 August. Available at: https://www.bbc.co.uk/news/uk-scotland-scotl and-business-58112938 (Accessed 28/07/2022).

Fraser, N. (1999). 'Social justice in the age of identity politics: Redistribution, recognition and Participation'. In: Ray, L. and Sayer, A. (eds.). *Culture and Economy after the Cultural Turn*. London: SAGE

Fraser, N. (2000). Rethinking recognition. New Left Review, 3: 107–120.

Friends of the Earth. (2021). *An Emergency Plan on Green Jobs for Young People.* London: Friends of the Earth.

Fthenakis, V.M. (2000). End-of-life management and recycling of PV modules. *Energy Policy*, 28: 1051–1058.

Gabbatiss, J., Graham, F., McSweeney, R. and Viglione, G. (2022). Factcheck: Is solar power a 'threat' to UK farmland? *Carbon Brief*, 25 August. Available at: https://www.carbonbrief.org/factcheck-is-solar-power-a-threat-to-uk-farmland/ (Accessed 09/12/2022).

Galgósczi, B. (2014). The long and winding road from black to green: Decades of structural change in the Ruhr region. *Int. J. Labour Res.* 6: 217.

Galgósczi, B. (2020). Just transition on the ground: Challenges and opportunities for social dialogue. *European Journal of Industrial Relations*, 26(4): 367–382.

Gallop, P. (2021). The public is still paying, hydropower operators are still profiting. *Bankwatch*, 27 January. Available at: https://bankwatch.org/blog/the-public-is-still-paying-hydropower-operators-are-still-profiting (Accessed 28/07/2022).

Galvin, R. (2018). Trouble at the end of the line: Local activism and social acceptance in low-carbon electricity transmission in Lower Franconia, Germany. *Energy Research & Social Science*, 38; 114–126.

Galvin, R. (2020). Power, evil and resistance in social structure: A sociology for energy research in a climate emergency. *Energy Research & Social Science*, 71: 101361.

Garcés, I. and Alvarez, G. (2020). Water mining and extractivism of the Salar de Atacama, Chile. *Transactions on Ecology and the Environment*, 245: 189–199.

García, J.H., Cherry, T.L., Kallbekken, S. and Torvanger, A. (2016). Willingness to accept local wind energy development: Does the compensation mechanism matter? *Energy Policy*, 99: 165–173.

Garside, J. (2021). How the Queen came to own the seabed around Britain. *The Guardian*, 5 February. Available at: https://www.theguardian.com/environment/2021/feb/05/how-the-queen-came-to-own-the-seabed-around-britain (Accessed 28/07/2022).

Garvey, A., Norman, J.B., Büchs, M. and Barrett, J. (2022). A "spatially just" transition? A critical review of regional equity in decarbonisation pathways. *Energy Research & Social Science*, 88: 102630.

Geiger, R., Hannan, Y., Travia, W., Naboni, R., and Schlette, C. (2020). Composite wind turbine blade recycling – value creation through Industry 4.0 to enable circularity in repurposing of composites. *IOP Conf. Series: Materials Science and Engineering*, 942: 012016.

Generation Rent. (no date.) About renting: 13 million people in the UK rent from a private landlord – that's 1 in every 5 of us. *Generation Rent*. Available at: https://www.generationrent.org/about_renting (Accessed 28/07/2022).

George, S. (2022). Labour Party proposes £8bn 'national wealth fund' to invest in green industrial transition. *Edie*, 26 September. Available at: https://www.edie.net/labour-party-proposes-8bn-national-wealth-fund-to-invest-in-green-industrial-transition/ (Accessed 09/12/2022).

George, S. and Mace, M. (2022). Rishi Sunak announces £5bn windfall tax on fossil fuel giants to help households deal with energy price crisis. *Edie*, 26 May. Available at: https://www.edie.net/rishi-sunak-announces-5bn-windfall-tax-on-fossil-fuel-giants-to-help-households-deal-with-ene rgy-price-crisis/ (Accessed 28/07/2022).

Gibbons, S. (2014). *Gone with the Wind: Valuing the Visual Impacts of Wind Turbines through House Prices*. Spatial Economics Research Centre, SEC Discussion Paper 159.

Gibbs, E. and Kerr, E. (2022). Mobilizing solidarity in factory occupations: Activist responses to multinational plant closures. *Economic and Industrial Democracy*, 43(2): 612–633.

Giglio, E. (2021). Extractivism and its socio-environmental impact in South America. Overview of the "lithium triangle". *América Crítica* 5 (1): 47–53.

Giordano, C. (2022). British Gas owner Centrica profits increase five-fold to £1.34 billion as energy bills soar. *The Independent*, 28 July. Available at: https://www.independent.co.uk/news/business/british-gas-profits-centrica-energy-prices-b2133672.html (Accessed 29/04/2023).

Glasgow City Council. (2016). Glasgow's House Strategy, Factsheet 12/2016: Fuel Poverty. Glasgow: Glasgow City Council.

Global Energy Talent Index. (2021). Global Energy Talent Index, 2021 Report.

Glüsing, J., Hage, S., Jung, A., Klawitter, N. and Schultz, S. (2021). The dirty truth about clean technologies. *Spiegel International*, 4 November. Available at: https://www.spiegel.de/international/world/mining-the-pla net-to-death-the-dirty-truth-about-clean-technologies-a-696d7adf-35db-4844-80be-dbd1ab698fa3 (Accessed 28/07/2022).

Goddard, G. and Farrelly, M.A. (2018). Just transition management: Balancing just outcomes with just processes in Australian renewable energy transitions. *Applied Energy*, 225: 110–123.

Godfrey, G. (2022). Will I get £400 energy rebate if bills are included in my rent? *Sky News*, 4 July. Available at: https://news.sky.com/story/how-do-you-get-your-400-energy-bills-rebate-if-you-pay-landlord-for-utilit ies-12642590 (Accessed 18/03/2023).

Goodnough, A. (2010). For Cape Cod wind farm, new hurdle Is spiritual. *The New York Times*, 4 January. Available at: https://www.nytimes.com/2010/01/05/science/earth/05wind.html (Accessed 28/07/2022).

Gosden, E. (2022). Wind and solar boom will bring energy surplus. *The Times*, 7 May. Available at: https://www.thetimes.co.uk/article/wind-and-solar-boom-will-bring-energy-surplus-zplgq39rn (Accessed 28/07/2022).

Gosden, E. and Brown, D. (2021). Avro Energy sent money to firms run by owners. *The Times*, 24 September. Available at: https://www.thetimes.co.uk/article/avro-energy-sent-money-to-firms-run-by-owners-x328n9 5qj (Accessed 09/12/2022).

Gov.UK. (2020). Renting social housing. *UK Government*, 4 February. Available at: https://www.ethnicity-facts-figures.service.gov.uk/housing/social-housing/renting-from-a-local-authority-or-housing-association-soc ial-housing/latest (Accessed 28/07/2022).

Gov.UK. (2021a). Joint Statement: International Just Energy Transition Partnership. *UK Government*, 2 November. Available at: https://www.gov.uk/government/news/joint-statement-international-just-energy-tra nsition-partnership (Accessed 28/07/2022).

Gov.UK. (2021b). Press release: Rigorous new targets for green building revolution. *UK Government*, 19 January. Available at: https://www.gov.uk/government/news/rigorous-new-targets-for-green-building-revolut ion#:~:text=In per cent202019 per cent20the per cent20government per cent20introduced,a per cent20zero per cent20net per cent20emissions per cent20target. (Accessed 28/07/2022).

Gov.UK. (2022). Contracts for Difference Allocation, Round 4 Results. *UK Government*, 7 July. Available at: https://www.gov.uk/government/news/rigorous-new-targets-for-green-building-revolution#:~:text=In per cent202019 per cent20the per cent20government per cent20introduced,a per cent20zero per cent20net per cent20emissions per cent20target. (Accessed 28/07/2022).

Graham, E. (2022). Tax relief for oil and gas is trouble for UK bills and energy transition. *E3G*, 30 May. Available at: https://www.e3g.org/news/tax-relief-oil-gas-trouble-uk-household-bills-energy-transition/ (Accessed 28/07/2022).

Grant Thornton. (2021). Report concerning the governance arrangements for Bristol Energy. Bristol City Council, 15 September.

Greenfield, N. (2022). Lithium mining is leaving Chile's Indigenous communities high and dry (literally). *NRDC*, 26 April. Available at: https://www.nrdc.org/stories/lithium-mining-leaving-chiles-indigenous-comm unities-high-and-dry-literally (Accessed 28/07/2022).

Grenville, K. (2022). 37 of 40 most marginal constituencies being hit harder by gas crisis. *Energy & Climate Intelligence Unit,* 9 February. Available at: https://eciu.net/media/press-releases/2022/37-of-40-most-margi nal-constituencies-being-hit-harder-by-gas-crisis (Accessed 17/03/2022).

Grossman, E. (2019). France's Yellow Vests: Symptom of a chronic disease. *Political Insight,* 10(1): 30–34.

Groves, J. (2013). Cut the green crap! Cameron reveals his private view of energy taxation and orders ministers to dump the eco-charges adding £110-a-year to bills. *The Daily Mail,* 21 November. Available at: https:// www.dailymail.co.uk/news/article-2510936/Cut-green-c-p-Camerons- private-view-energy-taxation-horrify-environmental-campaigners.html (Accessed 28/07/2022).

Gudde, P., Oakes, J., Cochrane, P., Caldwell, N., and Bury, N. (2021). The role of UK local government in delivering on net zero carbon commitments: You've declared a Climate Emergency, so what's the plan? *Energy Policy,* 154: 112245.

Gurley, L.K. (2022). Shifting America to solar power is a gruelling, low-paid job. *Vice,* 27 June. Available at: https://www.vice.com/en/article/z34eyx/ shifting-america-to-solar-power-is-a-grueling-low-paid-job (Accessed 28/07/2022).

GWNET [Global Women's Network for the Energy Transition]. (2019). *Women for Sustainable Energy: Strategies to Foster Women's Talent for Transformational Change.* Vienna: GWNET.

Haf, S. and Robison, R. (2020). How local authorities can encourage citizen participation in energy transitions. London: UK Energy Research Centre. Available at: https://ukerc.ac.uk/publications/how-local-authorities-can- encourage-citizen-participation/ (Accessed 26/07/2022).

Haggett, C. (2011). Understanding public responses to offshore wind power, *Energy Policy* 39(2): 503–510.

Hall, A. (2018). Of Brexit, the fracking lobby and the revolving door. *openDemocracy,* 28 June. Available at: https://www.opendemocracy.net/ en/opendemocracyuk/of-brexit-fracking-lobby-and-revolving-door/ (Accessed 09/12/2022).

Hall, D. (2019) *Benefits and Costs of Bringing Water, Energy Grid and Royal Mail into Public Ownership.* Discussion Paper. PSIRU.

Halliday, J. (2018). 28,000 jobs at risk in north of England over low-carbon economy. *The Guardian,* 22 October. Available at: https://www.theguard ian.com/uk-news/2018/oct/22/28000-jobs-at-risk-in-north-of-england- over-low-carbon-economy (Accessed 28/07/2022).

Hammond, G. (2022). UK housebuilders warn new rules and taxes will add £4.5bn to costs. *Financial Times,* 30 October. Available at: https://www. ft.com/content/35c26a46–63ba-465b-9d28-cbd90b105943 (Accessed 09/12/2022).

Hampton, P. (2015). *Workers and Trade Unions for Climate Solidarity: Tackling Climate Change in a Neoliberal World.* Abingdon-on-Thames: Routledge.

Hanna, T.M., Bozuwa, J. and Rao, R. (2022). *The Power of Community Utilities.* Climate and Community Project.

Hannon, M., Cairns, I., Braunholtz-Speight, T., Hardy, J, McLachlan, C., Mander, S. and Sharmina, M. (2022). Policies to Unlock UK Community Energy Finance. JEPO-D-22-01119, Available at SSRN: https://ssrn. com/abstract=4109070

Hargreaves, T., Hielscher, S., Seyfang, G. and Smith, A. (2013). Grassroots innovations in community energy: The role of intermediaries in niche development. *Global Environmental Change*, 23: 868–880.

Hausfather, Z. and Stein, A.. (2022). Blog: The Need for NEM Reforms and the False Argument About the Cost of Diablo Canyon. *The Breakthrough Institute*, 5 January. Available at: https://thebreakthrough.org/blog/nem-reforms-and-diablo-canyon (Accessed 28/07/2022).

Hayes, R. (2021). *Scottish Building Regulations.* London: Solar Energy UK.

Haynes, J. (2023). 'Begging bowl culture must end' – Conservative Mayor Andy Street blasts Levelling Up snub. *Birmingham Live*, 19 January. Available at: https://www.birminghammail.co.uk/news/midlands-news/begging-bowl-culture-must-end-26020771 (Accessed 17/03/2023).

Hazrati, M. and Heffron, R. (2021). Conceptualising restorative justice in the energy transition: Changing the perspectives of fossil fuels. Energy Research & Social Science, 78: 102115.

HBF [Home Builders Federation]. (2018). Industry now generating £38bn a year and supporting 700k jobs. *Home Builders Federation*, 24 July. Available at: https://www.showhouse.co.uk/news/housebuilding-generates-38bn-a-year-and-supports-700k-jobs/. (Accessed 28/07/2022).

Healy, N. and Barry, J. (2017). Politicizing energy justice and energy system transitions: Fossil fuel divestment and a "just transition". *Energy Policy*, 108, 451–459.

Healy, N., Stephens, J.C. and Malin, S.A. (2019). Embodied energy injustices: Unveiling and politicizing the transboundary harms of fossil fuel extractivism and fossil fuel supply chains. *Energy Research & Social Science*, 48: 219–234.

Heath, L. (2021). Housing supply falls to five-year low as pandemic takes toll. *Inside Housing*, 25 November. Available at: https://www.insidehousing.co.uk/news/news/housing-supply-falls-to-five-year-low-as-pandemic-takes-toll-73492 (Accessed 28/07/2022).

Heffernan, R., Heidegger, P. Köhler, G., Stock, A. and Wiese, K. (2022). *A Feminist European Green Deal: Towards an Ecological and Gender Just Transition.* Bonn: Friedrich-Ebert-Stiftung.

Heffron, R.J. and McCauley, D. (2014). Achieving sustainable supply chains through energy justice. *Applied Energy*, 123: 435–437.

Helm, D. (2017). *Cost of Energy Review*. London: UK Government.

Hendriks, R., Raphals, P., Bakker, K. and Christie, G. (2017). First Nations and hydropower: The case of British Coloumbia's Site C Dam project. *Social Science Research Council*, 21 November. Available at: https://items.ssrc.org/just-environments/first-nations-and-hydropower-the-case-of-british-columbias-site-c-dam-project/ (Accessed 26/07/2022).

Hernández, D. (2015). Sacrifice along the energy continuum: A call for energy justice. *Environmental Justice*, 8(4): 151–156.

Hess, D.J. (2018). Energy democracy and social movements: A multi-coalition perspectives on the politics of sustainability transitions. *Energy Research & Social Science*, 40: 177–189.

Hess, D.J., McKane, R.G. and Belletto, K. (2021). Advocating a just transition in Appalachia: Civil society and industrial change in a carbon-intensive region. *Energy Research & Social Science*, 75: 102004.

Hickel, J. (2019). The limits of clean energy. *Foreign Policy*, 6 September. Available at: https://foreignpolicy.com/2019/09/06/the-path-to-clean-energy-will-be-very-dirty-climate-change-renewables/ (Accessed 09/12/2022).

Hoicka, C.E., Lowitzch, J., Brisbois, M.C., Kumar, A. and Camargo, L.R. (2021). Implementing a just renewable energy transitions: Policy advice for transposing the new European rules for renewable energy communities. *Energy Policy*, 156: 112435.

Hope, M. (2017). Mapped: How fracking lobbyists from the UK and America have infiltrated Parliament. *DeSmog*, 26 January. Available at: https://www.desmog.com/2017/01/26/mapped-how-fracking-lobbyists-uk-and-america-infiltrate-parliament/ (Accessed 09/12/2022).

Horne, C., Kennedy, E..H. and Familia, T. (2021). Rooftop solar in the United States: Exploring trust, utility perceptions and adoption among California homeowners. *Energy Research & Social Science*, 82: 102308.

Horton, H. (2022). The Tory green consensus is breaking – this leadership contest could spell the end of net zero. *The Guardian*, 12 July. Available at: https://www.theguardian.com/commentisfree/2022/jul/12/tory-green-consensus-leadership-contest-net-zero-climate-sceptic (Accessed 28/07/2022).

Horton, H. and Taylor, M. (2022). It's all a bit cynical': the politicians behind the Tory attack on net zero agenda. *The Guardian*, 8 February. Available at: https://www.theguardian.com/environment/2022/feb/08/its-all-a-bit-cynical-the-politicians-behind-the-tory-attack-on-net-zero-agenda (Accessed 28/07/2022).

Huber, M. (2015). Theorizing energy geographies. *Geography Compass*, 9(6): 327–338.

Huber, M. (2022). *Climate Change as Class War: Building Socialism on a Warming Planet*. London: Verso.

Huber, M. and Stafford, F. (2022). In defense of the Tennessee Valley Authority. *Jacobin*, 4th April. Available at: https://jacobin.com/2022/04/new-deal-tennessee-valley-authority-electricity-public-utilities-renewab les-green-power (Accessed 25/07/2022).

Hughes, D. M. (2021). *Who Owns the Wind? Climate Crisis and the Hope of Renewable Energy*. London: Verso.

Hunold, C. and Young, I.M. (1998). Justice, democracy and hazardous siting. *Political Studies*, 46(1): 82–95.

Hunter, C. (2022). Medway Council told to refuse own solar panel plan for 'historic' headquarters. *KentOnline*, 22 June. Available at: https://www.ken tonline.co.uk/kent/news/council-told-to-refuse-its-solar-panel-plan-for-own-hq-269099/ (Accessed 09/12/2022).

Hutcheon, P. (2021). BiFab yard joy as nearly 300 jobs to be created for wind turbine foundation works. *Daily Record*, 16 April. Available at: https://www.dailyrecord.co.uk/news/politics/bifab-yard-joy-nearly-300–23925 097 (Accessed 28/07/2022).

Hutchinson, K. (2021). Solar insanity: An update on Cleve Hill. *The Faversham Eye*, 18 December. Available at: https://www.favershameye.co.uk/post/solar-insanity-an-update-on-cleve-hill (Accessed 25/07/2022).

IDMC [Internal Displacement Monitoring Centre]. (2017). *Case study series: Dam displacement*. Geneva: Internal Displacement Monitoring Centre. Available at: https://www.internal-displacement.org/publications/case-study-series-dam-displacement (Accessed 26/07/2022).

IEA [International Energy Agency. (2021a) *The Role of Critical Minerals in Clean Energy Transitions*. Paris: International Energy Agency.

IEA. (2021b). *Hydropower Special Market Report: Analysis and Forecast to 2030*. Paris: International Energy Agency.

IEA. (2022a). *Renewable Energy Market Update – May 2022*. Paris: International Energy Agency.

IEA. (2022b). Multiple benefits of energy efficiency: Health and wellbeing. *International Energy Agency*. Available at: https://www.iea.org/reports/multiple-benefits-of-energy-efficiency/health-and-wellbeing (Accessed 28/07/2022).

IEA. (2022c). *World Energy Employment*. Paris: International Energy Agency.

IHA [International Hydropower Association]. (2020). *2020 Hydropower Status Report: Sector trends and insights*. Available at: https://hydropower-ass ets.s3.eu-west-2.amazonaws.com/publications-docs/2020_hydropower_st atus_report.pdf (Accessed 26/07/2022).

IHA. (2021a). 2021 Hydropower Status Report underscores need for rapid growth to achieve net zero. *International Hydropower Association*, 11 June. Available at: https://www.hydropower.org/news/2021-hydropower-status-report-underscores-need-for-rapid-growth-to-achieve-net-zero (Accessed 26/07/2022).

IHA. (2021b). Analysis: Hydropower's flexibility integral to reducing reliance on coal. *International Hydropower Association*, 1 February. Available at: https://www.hydropower.org/blog/analysis-hydropowers-flexibility-integral-to-reducing-reliance-on-coal (Accessed 26/07/2022).

IHA. (2022). *2022 Hydropower Status Report: Sector trends and insights.* Available at: https://assets-global.website-files.com/5f749e4b9399c80b5e421384/62d95e9c1d2120ce0b891efc_IHA per cent20Hydropower per cent20Status per cent20Report per cent202022.pdf (Accessed 26/07/2022).

Illinois Government. (2021). Press release: Gov. Pritzker Signs Transformative Legislation Establishing Illinois as a National Leader on Climate Action. *Illinois Government*, 15 September. Available at: https://www.illinois.gov/news/press-release.23893.html (Accessed 15/09/2021).

ILO. [International Labor Organization]. (2015). ILO adopts guidelines on sustainable development, decent work and green jobs. *International Labor Organization*, 5 November. Available at: https://www.ilo.org/global/topics/green-jobs/news/WCMS_422575/lang—en/index.htm (Accessed 28/07/2022).

ILO. (2019). Boosting Skills for a Just Transition and the Future of Work – 6 June, 2019. *International Labor Organization*, no date. Available at: https://www.ilo.org/skills/areas/skills-training-for-poverty-reduction/WCMS_679547/lang—en/index.htm (Accessed 28/07/2022).

ILO. (2021) Decent work. *International Labor Organization*, no date. Available at: https://www.ilo.org/global/topics/decent-work/lang—en/index.htm (Accessed 27/07/2022).

ILO. (2022). ILO welcomes G7 call to make a just transition to a green economy happen. *International Labor Organization*, 24 May. Available at: https://www.ilo.org/global/about-the-ilo/newsroom/news/WCMS_846303/lang--en/index.htm(Accessed 18/03/2023).

Inglis, T. (2022). Three wind turbines and solar farm could be developed in Irvine. *Daily Record*, 22 February. Available at: https://www.dailyrecord.co.uk/ayrshire/three-wind-turbines-solar-farm-26290792 (Accessed 15/12/2022).

International Olympic Committee. (2021). London 2012: a legacy that keeps giving. *International Olympic Committee*, 13 April. Available at: https://olympics.com/ioc/news/london-2012-a-legacy-that-keeps-giving (Accessed 28/07/2022).

International Rivers. (2022). Sustainable Energy for all: making a case for community-scale micro-hydro as the solution. *International Rivers*, 4 February. Available at: https://www.internationalrivers.org/news/sustainable-energy-for-all-making-a-case-for-community-scale-micro-hydro-as-the-solution/ (Accessed 28/07/2022).

IPCC [Intergovernmental Panel on Climate Change]. (2022a). *Climate Change 2022: Impacts, Adaptation and Vulnerability*. Sixth Assessment Report: Working Group II.

IPCC. (2022b). *Climate Change 2022: Mitigation of Climate Change*. Sixth

IREC [Interstate Renewable Energy Council]. (2021). *National Solar Jobs Census 2020*. Albany, NY: Interstate Renewable Energy Council.

IRENA [International Renewable Energy Agency]. (2016). *End-of-Life Management: Solar Photovoltaic Panels*. Abu Dhabi: International Renewable Energy Agency.

IRENA. (2019). *Renewable Energy: A Gender Perspective*. Abu Dhabi: International Renewable Energy Agency.

IRENA. (2020a) *Wind Energy: A Gender Perspective*. Abu Dhabi: International Renewable Energy Agency.

IRENA. (2020b) *Renewable Energy and Jobs: Annual Review 2019*. Abu Dhabi: International Renewable Energy Agency.

IRENA. (2021a). *Renewable Energy Statistics, 2021*. Abu Dhabi: International Renewable Energy Agency.

IRENA. (2021b). *Renewable Power Generation Costs in 2020*. Abu Dhabi: International Renewable Energy Agency.

IRENA. (2021c). *Renewable energy and Jobs: Annual Review 2020*. Abu Dhabi: International Renewable Energy Agency.

IRENA. (2022). *Renewable Energy Statistics, 2022*. Abu Dhabi: International Renewable Energy Agency.

ITUC [International Trade Unions Congress] (2012). *Growing Green and Decent Jobs*. Brussels: ITUC.

ITUC. (2020). Scorecards: NDCs – #JustTransition for #ClimateAmbition. *International Trade Unions Congress*, 18 September. Available at: https://www.ituc-csi.org/scorecards-ndcs (Accessed 28/07/2022).

Jack, S. (2022). Most only have £500 of savings says Lloyds boss. *BBC News*, 6 July. Available at: https://www.bbc.co.uk/news/business-62057 301 (Accessed 27/07/2022).

Jackson, D.Z. (2010). The wind of change. *The Boston Globe*, 1 May. Available at: http://archive.boston.com/bostonglobe/editorial_opinion/oped/artic les/2010/05/01/the_winds_of_change/ (Accessed 28/07/2022).

Jeliazkov, G., Morrison, R. and Evans, M. (2020). OFFSHORE: Oil and gas workers' views on industry conditions and the energy transition. Platform, Friends of the Earth Scotland, and Greenpeace. Available at: https://pla tformlondon.org/p-publications/offshore-oil-and-gas-workers-views/ (Accessed: 25/07/2022).

Jenkins, K. (2019). Implementing just transition after COP24. *Climate Strategies Policy Brief*, January 2019. Available at: https://climatestrategies. org/wp-content/uploads/2019/01/Implementing-Just-Transition-after-COP24_FINAL.pdf (Accessed 25/07/2022).

Jenkins, K., McCauley, D., Heffron, R., Stephans, H. and Rehner, R. (2016). Energy justice: A conceptual review. *Energy Research & Social Science*, 11: 174–182.

Jennings, E. (2022). Twitter, 21 June. Available at: https://twitter.com/Ed_Jennings/status/1539327454036500480 (Accessed 28/07/2022).

Jensen, J.P. (2019). Evaluating the environmental impacts of recycling wind turbines. *Wind Energy*, 22: 316–326.

Jerez, B., Garcés, I. and Torres, R. (2021). Lithium extractivism and water injustices in the Salar de Atacama, Chile: The colonial shadow of green electromobility. *Political Geography*, 87: 102382.

Johnson, K., Kerr, S. and Side, J. (2013). Marine renewables and coastal communities: Experiences from the offshore oil industry in the 1970s and their relevance to marine renewables in the 2010s. *Marine Policy*, 38: 491–499.

Jones, C. (2021). As clean energy jobs grow, women and Black workers are at risk of being left behind. *USA Today*, 9 September. Available at: https://eu.usatoday.com/story/money/2021/09/09/clean-energy-jobs-women-black-hispanic-workers-report/5776787001/ (Accessed 18/03/2023).

Jones, N., Marks, R., Ramirez, R. and Ríos-Vargas, M. (2021). 2020 Census illuminate racial and ethnic composition of the country. *United States Census Bureau*, 12 August. Available at: https://www.census.gov/library/stories/2021/08/improved-race-ethnicity-measures-reveal-united-states-population-much-more-multiracial.html (Accessed 28/07/2022).

Just Transition Research Collaborative (2018). *Mapping Just Transition(s) to a Low-Carbon World*. UNRISD. Available at: https://www.uncclearn.org/wp-content/uploads/library/report-jtrc-2018.pdf (Accessed 25/07/2022).

Kalkbrenner, B.J. and Roosen, J. (2016). Citizens' willingness to participate in local renewable energy projects: The role of community and trust in Germany. *Energy Research & Social Science*, 13: 60–70.

Kaswan, A. (2021). 'Distributive environmental justice'. In Coolsaet, B. (ed.). *Environmental Justice: Key Ideas*. London: Earthscan for Routledge.

Katwala, A. (2018). The spiralling environmental cost of our lithium battery addiction. *Wired*, 5 August. Available at: https://www.wired.co.uk/article/lithium-batteries-environment-impact (Accessed 28/07/2022).

Katz, C. (2020). The batteries that could make fossil fuels obsolete. *Yale Environment 360*, 15 December. Available at: https://e360.yale.edu/features/in-boost-for-renewables-grid-scale-battery-storage-is-on-the-rise (Accessed 17/03/2023).

Kawuk, C. (2021). The road to a net-zero economy requires building girls' green skills for green jobs. *The Brookings Institute*, 1 March. Available at: https://www.brookings.edu/blog/education-plus-development/2021/03/01/the-road-to-a-net-zero-economy-requires-building-girls-green-skills-for-green-jobs/ (Accessed 27/07/2022).

Keating, C. (2022). 'Accelerate now': Why calls for green skills training are getting louder. *Business Green*, 8 August. Available at: https://www.busine ssgreen.com/news-analysis/4054396/accelerate-calls-green-skills-train ing-getting-louder (Accessed 09/12/2022).

Kelly, A. (2019). Apple and Google named in US lawsuit over Congolese child cobalt mining deaths. *The Guardian*, 16 December. Available at: https://www.theguardian.com/global-development/2019/dec/16/ apple-and-google-named-in-us-lawsuit-over-congolese-child-cobalt-min ing-deaths (Accessed 28/07/2022).

Kelly, M. (2021). FT 1000: the fifth annual list of Europe's fastest-growing companies. *Financial Times*, 2 March. Available at: https://www.ft.com/ content/8b37a92b-15e6-4b9c-8427-315a8b5f4332 (Accessed 09/12/ 2022).

Kemfert, C., Opitz, P., Traber, T. and Handrich, L. (2015). Deep decarbonization in Germany: A macro-analysis of economic and political challenges of the 'Energiewende' (Energy Transition). DIW Berlin: Politikberatung Kompakt, 93. Available at: https://www.econstor. eu/handle/10419/108671 (Accessed 26/07/2022).

Kempton, W., Firestone, J., Lilley, J., Rouleau, T. and Whitaker, P. (2005). The offshore wind power debate: Views from Cape Cod. *Coastal Management*, 33(2): 119–149.

Kennedy, Jr, R.F. (2005). An ill wind off Cape Cod. *The New York Times*, 16 December. Available at: https://www.nytimes.com/2005/12/16/opin ion/an-ill-wind-off-cape-cod.html (Accessed 28/07/2022).

Kennedy, S. (2021). Revealed: Fossil fuel companies lobby UK government for gas 'compromise' ahead of COP26. *Channel 4 News*, 7 July. Available at: https://www.channel4.com/news/revealed-fossil-fuel-companies- lobby-uk-government-for-gas-compromise-ahead-of-cop26 (Accessed 09/12/2022).

Kijewski, L. (2022). Portuguese villagers fear hunt for lithium will destroy their livelihoods. *Politico*, 27 April. https://www.politico.eu/article/ portugal-village-fear-hunt-lithium-destroy-livelihood/ (Accessed 28/07/ 2022).

Kim, E.S. and Chung, J.B. (2019). The memory of place disruption, senses and local opposition to Korean wind farms. *Energy Policy*, 131: 43–52.

Kingsbury, D.V. (2021). 'Green' extractivism and the limits of energy transitions: Lithium, sacrifice and maldevelopment of the Americas. *SFS: Georgetown Journal of International Affairs*, 20 July. Available at: https:// gjia.georgetown.edu/2021/07/20/green-extractivism-and-the-limits-of- energy-transitions-lithium-sacrifice-and-maldevelopment-in-the-ameri cas/ (Accessed 28/07/2022).

Kinniburgh, C. (2019). Climate politics after the Yellow Vests. *Dissent*, 66(2): 115–125.

Klæboe, R. and Sundfør, H.B. (2016). Windmill noise annoyance, visual aesthetics, and attitudes towards renewable energy sources. *Int J Environ Res Public Health*, 13(8): 746.

Klein, N. (2019). *On Fire: The Burning case for a Green New Deal*. London: Allen Lane.

Klessmann, C. and Tiedemann, S. (2017). Germany's first renewables auctions are a success, but new rules are upsetting the market. *Energypost. eu*, 27 June. Available at: https://energypost.eu/germanys-first-renewables-auctions-are-a-success-but-new-rules-are-upsetting-the-market (Accessed 28/07/2022).

Knox, S., Hannon, M., Stewart, F. and Ford, R. (2022). The (in)justices of smart local energy systems: A systematic review, integrated framework and future research agendas. *Energy Research & Social Science*, 83: 102333.

Knuth, S. (2021). Rentiers of the low-carbon economy? Renewable energy's extractive fiscal geographies. *Environment and Planning A: Economy and Space*, OnlineFirst.

Kohler, B. (1996). Sustainable development: A labor view. Presentation at 1996 Persistent Organic Pollutants Conference, 5 December 1996, University of Chicago. Available at: https://www.sdearthtimes.com/et0597/et0597s4.html (Accessed 16/03/2023).

Kolinjivadi, V. and Kothari, A. (2020). No harm is still harm there: The Green New Deal and the Global South. *Jamhoor*, 20 May. Available at: https://www.jamhoor.org/read/2020/5/20/no-harm-here-is-still-harm-there-looking-at-the-green-new-deal-from-the-global-south (Accessed 28/07/2022).

Kooij, H-J., Oteman, M., Veenman, S., Sperling, K., Magnusson, D., Palm, J. and Hvelplund, F. (2018). Between grassroots and treetops: Community power and institutional dependence in the renewable energy sector in Denmark, Sweden and the Netherlands. *Energy Research & Social Science*, 37: 52–64.

Krebel, L., Stirling, A., van Lerven, F. and Arnold, S. (2020). *Building a Green Stimulus for Covid-19: A Recovery Plan for a Greener, Fairer Future*. London: New Economics Foundation.

Kumar, A. and Turner, B. (2020). 'Sociomaterial solar waste: Afterlives and lives after of small solar'. In: Bombaerts, G., Jenkins, K., Sanusi, Y.A. and Guoyu, W. (eds.), *Energy Justice Across Borders*. Berlin: Springer.

Kwan, J. (2020). Your old electronics are poisoning people at this toxic dump in Ghana. *Wired,* 26 November. Available at: https://www.wired.co.uk/article/ghana-ewaste-dump-electronics (Accessed 28/07/2022).

La Sicilia. (2019). La scalata di Vito Nicastri da elettricista a "signore del vento". *La Sicilia*, 18 April. Available at: https://www.lasicilia.it/news/trapani/236743/la-scalata-di-vito-nicastri-da-elettricista-a-signore-del-vento.html (Accessed 28/07/2022).

Lacey-Barnacle, M. (2020) Proximities of energy justice: Contesting community energy and austerity in England. *Energy Research & Social Science*, 69: 101713.

Lacey-Barnacle, M. and Bird, C. (2018). Intermediating energy justice? The role of intermediaries in the civic energy sector in a time of austerity. *Applied Energy*, 226: 71–81.

Ladenburg, J. and Dubgaard, A. (2007). Willingness to pay for reduced visual disamenities from offshore wind farms in Denmark. *Energy Policy*, 35(8): 4059–4071.

Larkins, M. (2021). 'Keeping it local: The continued relevance of place-based studies for environmental justice research and praxis. In: Ryder, S., Powlen, K., Laituri, M., Malin, S.A., Sbicca, J. and Stevis, D. *Environmental Justice in the Anthropocene: From (Un)just Presents to Just Futures*. London: Earthscan.

Lawrence, F. and McSweeey, E. (2018). Migrants building £2.6bn windfarm paid fraction of minimum wage. *The Guardian*, 21 October. Available at: https://www.theguardian.com/uk-news/2018/oct/21/migrants-building-beatrice-windfarm-paid-fraction-of-minimum-wage (Accessed 28/07/2022).

Lawrence, R. (2014). Internal colonisation and Indigenous resource sovereignty: Wind power developments on traditional Saami lands. *Environment and Planning D: Society and Space*, 32: 1036–1053.

Lawson, A. (2022). Energy firms accused of profiteering with 'horrendous rates' for care homes. *The Guardian*, 16 November. Available at: https://www.theguardian.com/business/2022/nov/16/energy-firms-accused-profiteering-horrendous-rates-care-homes (Accessed 09/12/2022).

Lees, A.C., Peres, C.A., Fearnside, P.M., Schneider, M. and Zuanon, J.A.S. (2016). Hydropower and the future of Amazonian biodiversity. *Biodivers. Conserv.*, 25: 451–466.

Leffel, B. (2022). Climate consultants and complementarity: Local procurement, green industry and decarbonization in Australia, Singapore, and the United States. *Energy Research & Social Science*, 88: 102635.

Lennon, B., Dunphy, N., Gaffney, C., Revez, A., Mullally, G. and O'Connor, P. (2020). Citizen or consumer? Reconsidering energy citizenship. *Journal of Environmental policy and Planning*, 22(2): 184–197.

Lennon, M. (2021). Energy transitions in a time of intersecting precarities: From reductive environmentalism to antiracist praxis. *Energy Research & Social Science*, 73: 101930.

Lennon, M. and Rogers, D. (2021). Decentralizing energy. *American Anthropologist*. Available at: https://www.americananthropologist.org/deprovincializing-development-series/decentralizing-energy (Accessed 28/07/2022).

Leonhardt, R., Noble, B., Poelzer, G., Fitzpatrick, P., Belcher, K. and Holdmann, G. (2022). Advancing local energy transitions: A global review of government instruments supporting community energy. *Energy Research & Social Science*, 83: 102350.

Lerner, S. (2010). *Sacrifice Zones: The Front Lines of Toxic Chemical Exposure in the United States*. Cambridge, MA: MIT Press.

Levenda, A.M., Behrsin, I. and Disano, F. (2021). Renewable energy for whom? A global systematic review of the environmental justice implications of renewable energy technologies. *Energy Research & Social Science*, 71: 101837.

Lewis, J., Hernández, D. and Geronimus, A.T. (2019). Energy efficiency as energy justice: Addressing racial inequalities through investments in people and places. *Energy Efficiency*, 13(3): 419–432.

LGA [Local Government Association]. (2020). *Local Green Jobs: Accelerating a Sustainable Economic Recovery*. London: Local Government Association.

LGA. (2021). COP26 – Councils trusted most to lead the fight against climate change – new survey finds. *Local Government Association*, 11 November. Available at: https://www.local.gov.uk/about/news/cop26-councils-trusted-most-lead-fight-against-climate-change-new-survey-finds (Accessed 28/07/2022).

Liddell, C., Morris, C., Gray, B., Czerwinska, A. and Thomas, B. (2016). Excess winter mortality associated with Alzheimer's Disease and related dementias in the UK: A case for energy justice. *Energy Research & Social Science*, 11: 256–262.

Liebe, U., Bartczak, A. and Meyerhoff, J. (2017). A turbine is not only a turbine: The role of social context and fairness characteristics for the local acceptance of wind power. *Energy Policy*, 107: 300–308.

Lienhoop, N. (2018). Acceptance of wind energy and the role of financial and procedural participation: An investigation with focus groups and choice experiments. *Energy Policy*, 118: 97–105.

Lieu, J., Sorman, A.H., Johnson, O.W., Virla, L.D. and Resurrecion, B.P. (2020). Three sides to every story: Gender perspectives in energy transition pathways in Canada, Kenya and Spain. *Energy Research & Social Science*, 68: 101550.

Liu, W., Agusdinata, D.B. and Myint, S.W. (2019). Spatiotemporal patterns of lithium mining and environmental degradation in the Atacama Salt Flat, Chile. *Int. J. Appl. Earth Obs. Geoinformation*, 80: 145–156.

Liu, W. and Agusdinata, D.B. (2020). Interdependencies of lithium mining and communities sustainability in Salar de Atacama, Chile. *Journal of Cleaner Production*, 260: 120838.

Livsey, A. (2020). Lex in depth: The $900bn cost of 'stranded energy assets'. *Financial Times*, 3 February. Available at: https://www.ft.com/content/95efca74-4299-11ea-a43a-c4b328d9061c (Accessed 26/07/2022).

Louie, E.P. and Pearce, J.M. (2016). Retraining investment for U.S. transition from coal to solar photovoltaic employment. *Energy Economics*, 57: 295–302.

Lovins, A.B. (1976). Energy strategy: The road not taken? *Foreign Affairs*, October 1976. Available at: https://energyhistory.yale.edu/library-item/amory-lovins-energy-strategy-road-not-taken-foreign-affairs-1976 (Accessed 26/07/2022).

Lovins, A.B. (1977). Resilience in energy strategy. *New York Times*, 24 July. Available at: https://www.nytimes.com/1977/07/24/archives/resilience-in-energy-strategy.html (Accessed 09/12/2022).

Lucas Aerospace Combine Shop Steward Committee. (1976). *LUCAS: An Alternative Plan*. Nottingham: Russell Press Ltd.

Luke, N., Zabin, C., Velasco, D. and Collier, R. (2017). *Diversity in California's Clean Energy Workforce: Access to Jobs for Disadvantaged Workers in Renewable Energy Construction*. Center for Labor Research and Education, University of California, Berkeley.

Luo, Q., Fang, G., Ye, J., Yan, M. and Lu, C. (2020). Country Evaluation for China's hydropower investment in the Belt and Road Initiative nations. *Sustainability* 12, 8281.

Lybrand, H. and Subramaniam, T. (2021). Fact-checking the Texas energy-failure blame game. *CNN*, 19 February. Available at: https://edition.cnn.com/2021/02/19/politics/texas-energy-outage-wind-turbine-blame-green-energy-fact-check/index.html (Accessed 28/07/2022).

Malloy, K. (2021). UK on track to become Europe's biggest e-waste contributor. *Resource*, 21 October. Available at: https://resource.co/article/uk-track-become-europe-s-biggest-e-waste-contributor (Accessed 28/07/2022).

Malm, A. (2016). *Fossil Capital: The Rise of Steam Powerr and the Roots of Global Warming*. London: Verso.

Mang-Benza, C. (2021). Many shades of pink in the energy transition: Seeing women in energy extraction, production, distribution and consumption. *Energy Research & Social Science*, 73: 101901.

Manzo, L. and Perkins, D. (2006). Finding common ground: the importance of place attachment to community participation in planning. *Journal of Planning Literature*, 20: 335–350.

Marchand, J. (2012). Local labor market impacts of energy boom-bust-boom in Western Canada. *J. Urban. Econ*, 71: 165–174.

Marczinkowski, H.M., Østergaard, P.A. and Djørup, S.R. (2019). Transitioning island energy systems: Local conditions, development phases, and renewable energy integration. *Energies*, 12: 3483.

Marriott, J. and Macalister, T. (2021). *Crude Britannia: How Oil Shaped a Nation*. London: Pluto Press.

Marshall, J. (2021). *Bills, Bills, Bills: How Rising Energy Costs Will Impact More Than Others and What the Government Can Do About It*. London: Resolution Foundation.

Martin, C. (2020). Wind turbine blades can't be recycled, so they're piling up in landfills. *Bloomberg*, 5 February. Available at: https://www.bloomb erg.com/news/features/2020-02-05/wind-turbine-blades-can-t-be-recyc led-so-they-re-piling-up-in-landfills#xj4y7vzkg (Accessed 28/07/2022).

Martin, D. (2022). Rishi Sunak: We won't lose our best farmland to solar panels. *The Telegraph*, 18 August. Available at: https://www.telegraph. co.uk/politics/2022/08/18/rishi-sunak-wont-lose-best-farmland-solar-panels/ (Accessed 09/12/2022).

Martin, M. and Islar, M. (2021). The 'end of the world' vs. the 'end of the month': understanding social resistance to sustainability transition agendas, a lesson from the Yellow Vests in France. *Sustainability Science*, 16: 601–614.

Martínez Rodríguez, R. (2021). Just transition: Time for a rethink? *Green European Journal*, 10 February. Available at: https://www.greeneuropean journal.eu/just-transition-time-for-a-rethink/ (Accessed 28/07/2022).

Martinson, K., Stanczyk, A. and Eyster, L. (2010). *Low-Skill Workers' Access to Quality Green Jobs*. The Urban Institute, Brief 13. Washington DC: Urban Institute.

Massey, D. (1991). A global sense of place. *Marxism Today*, June 1991.

Mathers, M. (2022). Energy bills to be burned in nationwide cost of living protests. *The Independent*, 1 October. Available at: https://www.independ ent.co.uk/news/uk/home-news/energy-bills-burned-protests-uk-b2188 424.html (Accessed 09/12/2022).

Mavrokefalidis, D. (2022). Households to be 'offered £15k energy efficiency grants'. *Energy Live News*, 25 November. Available at: https://www.ene rgylivenews.com/2022/11/25/households-to-be-offered-15k-energy-eff iciency-grants/ (Accessed 09/12/2022).

Mayer, A. (2018). A just transition for coal miners? Community identity and support from local policy actors. *Environmental Innovation and Societal Transitions*, 28: 1–13.

Mayes, J. (2022). UK slaps 25% windfall tax on profits of oil and gas firms. *Bloomberg*, 26 May. Available at: https://www.bloomberg.com/news/ articles/2022-05-26/uk-slaps-25-windfall-tax-on-profits-of-oil-and-gas-companies (Accessed 09/12/2022).

Mayor of London. (2019). Consultation response – Future Homes Standard: changes to Part L and Part F of the Building Regulations for new dwellings. London: Mayor of London.

McCarthy, J. (2015). A socioecological fix to capitalist crisis and climate change? The possibilities and limits of renewable energy. *Environment and Planning A: Economy and Space*, 47(12): 2845–2502.

McCauley, D. (2018). *Energy Justice: Re-Balancing the Trilemma of Security, Poverty and Climate Change*. London: Palgrave Macmillan.

McCauley, D. and Heffron, R. (2018). Just transition: Integrating climate, energy and environmental justice. *Energy Policy*, 119: 1–7.

McClure, L.A., LeBlanc, W.G., Fernandez, C.A., Fleming, L.E., Lee, D.J., Moore, K.J. and Caban-Martinez, A.J. (2017). Green collar workers: An emerging workforce in the environmental sector. *J Occup Environ Med.*, 59(5): 440–445.

McCully, P. (2001). *Silenced Rivers: The Ecology and Politics of Large Dams*. London: Zed Books.

McGlynn, M. (2022). Retrofit scheme 'not enough' for low income, energy poor households. *Irish Examiner*, 9 February. Available at: https://www.irishexaminer.com/news/arid-40804071.html (Accessed 28/07/2022).

Mejía-Montero, A., Lane, M., van der Horst, D. and Jenkins, K.E.H. (2021). Grounding the energy justice lifecycle framework: An exploration of utility-scale wind power in Oaxaca, Mexico. *Energy Research & Social Science*, 75: 102017.

Merme, V., Ahlers, R. and Guta, J. (2014). Private equity, public affair: Hydropower financing in the Mekong Basin. Global Environmental Change, 24: 20–29.

Merrick, R. (2022). Grant Shapps attacks new onshore wind turbines as 'eyesore on the hills', exposing cabinet split. *The Independent*, 3 April. Available at: https://www.independent.co.uk/news/uk/politics/energy-strategy-wind-turbines-grant-shapps-b2049866.html (Accessed 09/12/2022).

Meyerhoff, J. (2013). Do turbines in the vicinity of respondent's residences influence choices among programmes for future wind power generation? The Journal of Choice Medicine, 7: 58–61.

Meyerhoff, J., Ohl, C. and Hartje, V. (2010). Landscape externalities from onshore wind power. *Energy Policy*, 38: 82–92.

MHCLG [Ministry for Housing, Communities and Local Government, UK Department for]. (2019). *English Housing Survey: Households Report, 2017-2018*. London: UK Government.

Milburn, K. (2022). Don't Pay took down Kwasi Kwarteng: The campaign posed an 'existential threat' to the energy sector. *Novara Media*, 18 October. Available at: https://novaramedia.com/2022/10/18/dont-pay-took-down-kwasi-kwarteng/ (Accessed 09/12/2022).

Mildenberger, M., Howe, P.D. and Miliaich, C. (2019). Households with solar installations are ideologically diverse and more politically active than their neighbours. *Nature Energy*, 4: 1033–1039.

Mills, S.B., Bessette, D. and Smith, H. (2019). Exploring landowners' post-construction changes in perceptions of wind energy in Michigan. *Land Use Policy*, 82: 754–762.

Milne, E. (1966). Hansard Archive, 20 October, Blyth Shipyard (Closure). Vol 734, cc5690–78.

Mirletz, H., Ovaitt, S., Sridhar, S. and Barnes, T.M. (2022). Circular economy priorities for photovoltaics in the energy transition. *PLoS ONE*, 17(9): e0274351.

Mnookin, S. (2017). How Robert F. Kennedy, Jr, distorted vaccine science. *Scientific American*, 11 January. Available at: https://www.scientificameri can.com/article/how-robert-f-kennedy-jr-distorted-vaccine-science1/ (Accessed 28/07/2022).

Moore, J. (2019). Voices: Jeremy Corbyn's plans to nationalise the energy grid are needless and impractical. *The Independent*, 16 May. Available at: https:// www.independent.co.uk/voices/corbyn-national-grid-energy-national ise-green-industrial-revolution-a8916711.html (Accessed 28/07/2022).

Morales, E. (2020) Twitter, 26 July. Available at: https://twitter.com/evoe spueblo/status/1287411226574823429

Morales Balcázar, R. (2021). 'Crisis y minería del litio en el salar de Atacama. La necesidad de una mirada desde la justicia climática.' In Morales Balcázar, R. (ed.). *Salares Andinos. Ecología de Saberes por la Protecciòn de Nuestros Salres y Humedales*. Observatorio Plurinacional de Salares Andinos.

Morby, A. (2022). Factory to build subsea cable from Sahara to Devon approved. *Construction Enquirer*, 19 August. Available at : https://www. constructionenquirer.com/2022/08/19/factory-to-build-subsea-cable-from-sahara-to-devon-approved/ (Accessed 17/03/2023).

Morena, E., Krause, D. and Stevis, D. (2020) (eds.) *Just Transitions: Social Justice in the Shift Towards a Low-Carbon World*. London: Pluto Press.

Morison, R. (2022). A third of UK homes seen falling into energy poverty by October. *Bloomberg*, 21 July. Available at: https://www.bloomberg.com/ news/articles/2022–07–21/a-third-of-uk-homes-seen-falling-into-ene rgy-poverty-by-october#xj4y7vzkg (Accessed 28/07/2022).

Morris, C. and Jungjohann, A. (2016). *Energy Democracy: Germany's ENERGIEWENDE to Renewables*. London: Palgrave Macmillan.

Morris, S. (2022). Cost of living: Boris Johnson criticised for 'out of touch' response to pensioner riding bus to keep warm. *Sky News*, 3 May. Available at: https://news.sky.com/story/cost-of-living-utterly-shameful-that-pen sioners-are-riding-buses-to-keep-warm-says-labour-12604615 (Accessed 28/07/2022).

Moss, T., Becker, S. and Naumann, M. (2015). Whose energy transition is it, anyway? Organisation and ownership of the Energiewende in villages, cities and regions. *Local Environment*, 20(12): 1547–1563.

Motavalli, J. (2021). The NIMBY threat to renewable energy. *The Sierra Club*, 20 September. Available at: https://www.sierraclub.org/sierra/2021–4-fall/feature/nimby-threat-renewable-energy / (Accessed 09/12/2022).

Mulvaney, D. (2013). Opening the black box of solar energy technologies: Exploring tensions between innovation and environmental justice. *Science as Culture*, 22(2): 230–237.

Mulvaney, D. (2019). *Solar Power: Innovation, Sustainability, and Environmental Justice*. Oakland: University of California Press.

Mulvaney, D. (2022). Battle over solar power in the Golden State. *Bulletin of the Atomic Scientists*, 3 June. Available at: https://thebulletin.org/2022/06/battle-over-solar-power-in-the-golden-state/ (Accessed 28/07/2022).

Mundaca, L., Busch, H. and Schwer, S. (2018). 'Successful'low-carbon energy transitions at the community level? An energy justice perspective. *Applied Energy*, 218: 292–303.

Munday, M., Bristow, G. and Cowell, R. (2011). Wind farms in rural areas: How far do community benefits from wind farms represent a local economic development opportunity? *Journal of Rural Studies*, 27: 1–12.

Muotka, P.H. (2020). The Saami Council addresses UN Special Rapporteur regarding Øyfjellet Wind AS. *Sámiráđđi*, 18 September. Available at: https://www.saamicouncil.net/news-archive/the-saami-council-addresses-u-special-rapporteur-regarding-the-oyfjellet-wind (Accessed 09/12/2022).

Muro, M., Tomer, A., Shivaram, R. and Kane, J.W. (2019). Advancing inclusion through clean energy jobs. *Brookings Institute*, 18 April. Available at: https://www.brookings.edu/research/advancing-inclusion-through-clean-energy-jobs/ (Accessed 27/07/2022).

Murphy, L.T. and Elimä, N. (2021). *In Broad Daylight: Uyghur Forced Labour and Global Solar Supply Chains*. Sheffield: Helena Kennedy Centre for International Justice, Sheffield Hallam University.

Musall, F.D. and Kuik, O. (2011). Local acceptance of renewable energy: A case study from southeast Germany, *Energy Policy*, 39(6): 3252–3260.

Nailer, A. (2022). Wind turbine folly: Anthony Nailer explains why wind farms are a terrible idea... . *UK Independence Party*, 13 January. Available at: https://www.ukip.org/wind-turbine-folly

National Energy Action. (2020). New ONS figures reveal cold homes death toll. *National Energy Action*, 27 November. Available at: https://www.nea.org.uk/news/271120–01/ (Accessed 28/07/2022).

National Grid (UK) (2021). Building a platform to lead the energy transition: 2020/21 Full Year Results Statement. *National Grid*, 20 May. Available at: https://www.nationalgrid.com/document/141786/download (Accessed 25/07/2022).

National Grid (UK) (2022). Britain's Electricity Explained: May 2022. London: National Grid.

Naumann, M. and Rudolph, D. (2020). Conceptualizing rural energy transitions: Energizing rural studies, ruralizing energy research. *Energy Policy*, 73: 97–104.

NEF [New Economics Foundation]. (2022). Poorest 10% of families will see energy costs increase by £724, A 7.5 times larger rise than the richest 10% of families. *New Economics Foundation*, 3 February. Available at: https://neweconomics.org/2022/02/poorest-10-of-families-will-see-energy-costs-increase-by-724-a-7-5-times-larger-rise-than-the-richest-10-of-families (Accessed 28/07/2022).

Newell, P. (2021). *Power Shift: The Global Political Economy of Energy Transitions.* Cambridge: Cambridge University Press.

Nicholls, J. (2020). Technological intrusion and communicative renewal: The case of two rural solar farm developments in the UK. Energy policy, 139: 111287.

Nilson, R.S. and Stedman, R.C. (2021). Are big and small solar separate things?: The importance of scale in public support for solar energy development in upstate New York. *Energy Research & Social Science*, 86: 102449.

Nilsson, C., Reidy, C.A., Dynesius, M. and Revenga, C. (2005). Fragmentation and flow regulation of the world's large river systems. *Science*, 308: 405–408.

Nolden, C. (2013). Governing community energy – Feed-in tariffs and the development of community wind energy schemes in the United Kingdom and Germany. *Energy Policy*, 63: 543–552.

Normann, S. (2021). Green colonialism in the Nordic context: Exploring South Saami representations of wind energy development. *J. Community Psychol.*, 49: 77–94.

NREL [National Renewable Energy Laboratory, USA]. (2021). What it takes to realize a circular economy for solar photovoltaic system materials. *NREL*, 2 April. Available at: https://www.nrel.gov/news/program/2021/what-it-takes-to-realize-a-circular-economy-for-solar-photovoltaic-system-materials.html (Accessed 15/12/2022).

Nussbaum, M.C. (2011). *Creating Capabilities: The Human Development Approach.* Cambridge, MA: Belknap Press.

Nuttall, P. (2022). Why can't the UK manage to insulate its homes? *The New Statesman*, 18 March. Available at: https://www.newstatesman.com/environment/2022/03/why-cant-the-uk-manage-to-insulate-its-homes (Accessed 28/07/2022).

O'Donnell, N. and Rinaldi, O. (2022). Veterans struggle to find work after military: "We still want to give the best of ourselves". *CNBC*, 23 February. Available at: https://www.cbsnews.com/news/veterans-jobs-american-corporate-partners/ (Accessed 28/07/2022).

O'Donoghue, A.J. (2021). Are children 'dying like dogs' in effort to build better batteries? *Deseret News*, 24 May. Available at: https://www.dese ret.com/utah/2021/5/23/22441889/our-children-are-dying-like-dogs-congo-slave-labor-cobalt-lawsuit-apple-tesla-human-rights-dell (Accessed 28/07/2022).

O'Grady, C. (2022). Scotland's billionaires are turning climate change into a trophy game. *The Atlantic*, 20 May. Available at: https://www.theatlan tic.com/science/archive/2022/05/scotland-climate-change-land-use/629 835/ (Accessed 28/07/2022).

O'Shaughnessy, E. (2022) Rooftop solar incentives remain effective for low- and moderate-income adoption. *Energy Policy*, 163: 112881.

O'Shaughnessy, E., Barbose, G., Wiser, R., Forrester, S. and Darghouth, N. (2020). Toward income equity in rooftop solar adoption: The impact of policies and business models. *Nature Energy*, 6: 84–91.

O'Sullivan, K., Golubchikov, O. and Mehmood, A. (2020). Uneven energy transitions: Understanding continued energy peripheralization in rural communities. *Energy Policy*, 138: 111288.

Octopus Energy. (2022). Introducing Octopus Fan Club: Local green energy. From our turbines for your home. Available at: https://octopus.energy/ octopus-fan-club/ (Accessed 28/07/2022).

Octopus Group. (2023). Octopus Group's financial position. *Octopus Group*, no date. Available at: https://octopusgroup.com/octopus-groups-financ ial-position/ (Accessed 18/03/2023).

Ofgem. (2019). Ofgem gives go-ahead to Orkney transmission link subject to conditions. *Ofgem*, 16 September. Available at: https://www.ofgem. gov.uk/publications/ofgem-gives-go-ahead-orkney-transmission-link-subj ect-conditions (Accessed 28/07.2022).

OIC [Orkney Islands Council]. (2017). *Orkney Hydrogen Strategy: The Hydrogen Islands, 2019– 2025*. Kirkwall: Orkney Islands Council. Available at: https://www.oref.co.uk/wp-content/uploads/2020/11/Hydrogen-Strategy.pdf

Old Sparky. (2020). Analysis: How £35 million of public money was 'lost' to Bristol Energy. *The Bristol Cable*, 21 May. Available at: https://thebrist olcable.org/2020/05/analysis-how-35-million-of-public-money-was-lost-to-bristol-energy/ (Accessed 28/07/2022).

Olenick, M., Flowers, M. and Diaz, V.J. (2015). US veterans and their unique issues: enhancing health care professional awareness. *Adv Med Educ Pract*, 6: 635–639.

Olson-Hazoun, S.K. (2018). 'Why are we being punished and they are being rewarded?': Views on renewable energy in fossil fuels-based communities of the U.S. West. *The Extractive Industries and Society*, 5: 366–374.

ONS [Office for National Statistics, UK]. (2018). UK private rented sector: 2018. *Office for National Statistics*, 18 January. Available at: https://www.ons.gov.uk/economy/inflationandpriceindices/articles/ukprivaterentedsector/2018 (Accessed 28/07/2022).

ONS. (2020). Living longer: changes in housing tenure over time. *Office for National Statistics*, 10 February. Available at: https://www.ons.gov.uk/peoplepopulationandcommunity/birthsdeathsandmarriages/ageing/articles/livinglonger/changesinhousingtenureovertime (Accessed 28/07/2022).

ONS. (2021). An overview of workers who were furloughed in the UK: An overview of workers who were furloughed in the UK: October 2021. *Office for National Statistics*, 1 October. Available at: https://www.ons.gov.uk/employmentandlabourmarket/peopleinwork/employmentandemployeetypes/articles/anoverviewofworkerswhowerefurloughedintheuk/october2021 (Accessed 09/12/2022).

ONS. (2022a). Average household income, UK: financial year ending 2021. *Office for National Statistics*, 28 March. Available at: https://www.ons.gov.uk/peoplepopulationandcommunity/personalandhouseholdfinances/incomeandwealth/bulletins/householddisposableincomeandinequality/financialyearending2021 (Accessed 28/07/2022).

ONS. (2022b). Energy prices and their effect on households *Office for National Statistics*, 1 February. Available at: https://www.ons.gov.uk/economy/inflationandpriceindices/articles/energypricesandtheireffectonhouseholds/2022-02-01 (Accessed 28/07/2022).

ONS. (2022c). Energy efficiency of housing in England and Wales: 2022. *Office for National Statistics*, 25 October. Available at: https://www.ons.gov.uk/peoplepopulationandcommunity/housing/articles/energyefficiencyofhousinginenglandandwales/2022 (Accessed 17/03/2023).

ONS. (2022d). Percentage of dwellings by main fuel type used for central heating in newly built dwellings, England and Wales, 2020 to 2022. *Office for National Statistics*, 30 September. Available at: https://www.ons.gov.uk/peoplepopulationandcommunity/housing/adhocs/15106percentageofdwellingsbymainfueltypeusedforcentralheatinginnewlybuiltdwellingsenglandandwales2020to2022 (Accessed 17/03/2023).

ONS. (2023). Housing, England and Wales: Census 2021. *Office for National Statistics*, 5 January. Available at: https://www.ons.gov.uk/peoplepopulationandcommunity/housing/bulletins/housingenglandandwales/census2021 (Accessed 17/03/2023).

Ottinger, G. (2013). The winds of change: Environment justice in energy transitions. *Science as Culture*, 22(2): 222–229.

Pai, S., Emmerling, J., Drouet, L., Zerriffi, H., Jewell, J. (2021). Meeting well-below 2C target would increase energy sector jobs globally. *One Earth*, 4: 1026–1036.

Palit, D. and Kumar, A. (2022). Drivers and barriers to rural electrification in India – A multi-stakeholder analysis. *Renewable and Sustainable Energy Reviews*, 166: 112663.

Parker, G., Pickard, J. and Giles, C. (2022). Liz Truss unveils £150bn UK energy plan but limits business support. *Financial Times*, 8 September. Available at: https://www.ft.com/content/984129f9-a133–468b-bc38-e8c4ec7386d6 (Accessed 09/12/2022).

Patrick, H. (2022). E.ON announces profits of more than £3bn amid cost of living crisis. *The Independent*, 10 August. Available at: https://www.independent.co.uk/tv/news/energy-eon-cost-living-bills-b2142236.html (Accessed 09/12/2022).

Paul, F.C. and Cumbers, A. (2021). The return of the local state? Failing neoliberalism, remunicipalisation, and the role of the state in advanced capitalism. *EPA: Economy and Space*, OnlineFirst.

Paul, H.K. and Gebrial, D. (eds.) (2021). *Perspectives on a Global Green New Deal*. London: Rosa Luxemburg Stiftung.

Pearse, R. and Bryant, G. (2022). Labour in transition: A value-theoretical approach to renewable energy labour. *Environment and Planning E: Nature and Space*, 5(4): 1872–1894.

Pearson, R. (2022). Twitter, 16 April. Available at: https://twitter.com/rosiep4/status/1515206362950742017?s=11&t=0a0phdIpG2w8AI2WHmUghg (Accessed 28/07/2022).

Pegels, A. and Lütkenhorst, W. (2014). Is Germany's energy transition a case of successful green industrial policy? Contrasting wind and solar PV. *Energy Policy*, 74: 522–534.

Pemberton, S., Fahmy, E., Sutton, E. and Bell, K. (2016). Navigating the stigmatised identities of poverty in austere times: Resisting and responding to narratives of personal failure. *Critical Social Policy*, 36(1): 21–37.

Penn, I. (2020). N.A.A.C.P. tells local chapters: Don't let energy industry manipulate you. *The New York Times*, 5 January. Available at: https://www.nytimes.com/2020/01/05/business/energy-environment/naacp-utility-donations.html (Accessed 28/07/2022).

Penner, A.M. (2008). Race and gender differences in wages: The role of occupational sorting at the point of hire. *The Sociological Quarterly*, 49(3): 597–614.

Perry, S. (2019). Reunion event marks tenth anniversary of Vestas wind turbine factory occupation. *On the Wight*, 25 July. Available at: https://onthewight.com/reunion-event-marks-tenth-anniversary-of-vestas-wind-turbine-factory-occupation/ (Accessed 28/07/2022).

Pettifor, A. (2019). *The Case for the Green New Deal*. London: Verso.

Phillips, A. (2022). National Grid's profits rise to £3.4bn amid soaring energy bills. *Sky News*, 19 May. Available at: https://news.sky.com/story/national-grids-profits-rise-to-3-4bn-amid-growing-energy-bills-12616 860 (Accessed 09/12/2022).

Phillips, J., Britchfield, C. and Guertler, P. (2022). *Home Energy Security Strategy: The Permanent Solution for Lower Bills*. London: E3G.

Pichler, M., Krenmayr, N., Maneka, D., Brand, U., Högelsberger, H. and Wissen, M. (2021). Beyond the jobs-versus-environment dilemma? Contested social-ecological transformations in the automotive industry. *Energy Research & Social Science*, 79: 102180.

Pickard, J. and Thomas, N. (2022). Taxpayers face additional £500mn bill for Bulb Energy bailout. *Financial Times*, 23 March. Available at: https://www.ft.com/content/30ecda9a-c223-4b8e-b700-0490d4e0f352 (Accessed 09/12/2022).

Pickrell, E. (2022). Russia-Ukraine war helps drive nickel prices, EV headaches. Forbes, 31 March. Available at: https://www.forbes.com/sites/uhenergy/2022/03/31/russia-ukraine-war-helps-drive-nickel-prices-ev-headaches/?sh=f45649557cd9 (Accessed 28/07/2022).

Piddington, J., Nicol, S., Garrett, H. and Custard, M. (2020). *The Housing Stock of the United Kingdom*. Watford: BRE Trust.

Pinker, A. (2020). *Just Transitions: A Comparative Perspective*. A report prepared for the Scottish Just Transition Commission. *Scottish Government*, 25 August. Available at: https://www.gov.scot/publications/transitions-comparative-perspective/ (Accessed: 25/07/2022).

Pipe, E. (2021). Funding secured for Bristol's first hydro-electric generator. *Bristol 24/7*, 11 February. Available at: https://www.bristol247.com/news-and-features/news/funding-secured-for-bristols-first-hydro-electric-generator/ (Accessed 28/07/2022).

Pitt, J. and Nolden, C. (2020). Post-subsidy solar PV business models to tackle fuel poverty in multi-occupancy social housing. *Energies*, 13: 4852.

Platform. (2014). 'Energy beyond neoliberalism', In: Hall, S., Massey, D. and Rustin, M. (eds.) (2015). *After Neoliberalism: The Kilburn Manifesto*. London: Lawrence Wishart Books.

Platform. (2022). Cambo & Viking Energy – The common wealth of wind in Shetland. *Platform*, 18 January. Available at: https://platformlondon.org/2022/01/18/cambo-viking-energy-the-common-wealth-of-wind-in-shetland/ (Accessed 28/07/2022).

Pohlmann, A. and Colell, A. (2020). Distributing power: Community energy movements claiming the grid in Berlin and Hamburg. *Utilities Policy*, 65: 101066.

Pollin, R. and Callaci, B. (2019). The economics of just transition: A framework for supporting fossil fuel-dependent workers. *Labor Studies Journal*, 44(2); 93–138.

Possible. (no date.). Energy local: Making energy work for you, 2016–2018. *Possible*. Available at: https://www.wearepossible.org/energy-local (Accessed 28/07/2022).

Postans, A. (2021). New report reveals 'full picture' of Bristol Energy fiasco. *Bristol 24/7*, 21 September. Available at: https://www.bristol247.com/news-and-features/news/new-report-reveals-full-picture-of-bristol-energy-fiasco/ (Accessed 28/07/2022).

Power, M., Newell, P., Baker, L., Bulkeley, H., Kirshner, J. and Smith, A. (2016). The political economy of energy transitions in Mozambique and South Africa: The role of rising powers. *Energy Research & Social Science*, 17: 10–19.

Power Technology. (2020). The world's longest power transmission lines. *Power Technology*, 29 January. Available at: https://www.power-technology.com/analysis/featurethe-worlds-longest-power-transmission-lines-4167964/#:~:text=The per cent202 per cent2C543km per cent2Dlong per cent20Belo,to per cent20Rio per cent20de per cent20Janeiro per cent2C per cent20Brazil (Accessed 28/07/2022).

Property Industry Alliance. (2017). Property Data Report 2017. Property Industry Alliance. Available at: https://bpf.org.uk/media/3278/bpf-pia-property-report-2017-final.pdf (Accessed 08/12/2022).

Prynn, J. (2012). The Olympics boom created 100,000 jobs in London. *Evening Standard*, 17 October. Available at: https://www.standard.co.uk/news/london/the-olympics-boom-created-100-000-jobs-in-london-8214954.html (Accessed 18/03/2023).

Pulido, L. (2000). Rethinking environmental racism: white privilege and urban development in Southern California. *Annals of the Association of American Geographers*. 90(1): 12–40.

Putnam, T. and Brown, D. (2021). Grassroots retrofit: Community governance and residential energy transitions in the United Kingdom. *Energy Research & Social Science*, 78: 102102.

Rai-Roche, S. (2022). The world installed 174GW of solar in 2021 and is on track to deploy 260GW by end of 2022 – IEA. *PV Tech*, 25 October. Available at: https://www.pv-tech.org/the-world-installed-174gw-of-solar-in-2021-and-is-on-track-to-deploy-260gw-by-end-of-2022-iea/ (Accessed 09/12/2022).

Ramirez, J. and Böhm, S. (2021). Transactional colonialism in wind energy investments: Energy injustices against vulnerable people in the Isthmus of Tehuantepec. *Energy Research & Social Science*, 78: 102135.

Rampion 2. (2021). Statement of Community Consultation. Newhaven: Rampion.

Rampion Offshore Wind. (2022). First anniversary of Rampion Fund: Supporting over 1 million people in Sussex. *Rampion Offshore Wind*, 28 April. Available at: https://www.rampionoffshore.com/news/news-eve nts/fifth-anniversary-of-rampion-fund-supporting-over-1-million-peo ple-in-sussex/ (Accessed 28/07/2022).

Rand, J. and Hoen, B. (2017). Thirty years of North American wind energy acceptance research: What have we learned? *Energy* Research & Social Science, 29: 135–148.

Re-Wind Design. (2018). *Re-Wind Design Atlas*. Re-Wind Design Network.

Read, S. (2022). Solar panel sales boom as energy bills soar. *BBC News*, 28 August. Available at: https://www.bbc.co.uk/news/business-62524031 (Accessed 09/12/2022).

Reames, T.G. (2019). Addressing with state-level residential solar energy policies. *Renewable Energy Policy Initiative*, September 2019. Available at: https://closup.umich.edu/research/addressing-equity-state-level-resi dential-solar-energy-policies (Accessed 17/03/2023).

Reames, T.G. (2020). Distributional disparities in residential rooftop solar potential and penetration in four cities in the United States. *Energy Research & Social Science*, 96: 101612.

Reed, M. (2016). UKIP energy: promoting distrust and climate scepticism on social media. *LSE British Politics and Policy*, 27 June. Available at: https:// blogs.lse.ac.uk/politicsandpolicy/ukip-climate-scepticism/ (Accessed 28/ 07/2022).

Reid, L.W. and Rubin, B.A. (2003). Integrating economic dualism and labor market segmentation: the effects of race, gender, and structural location on earnings, 1974–2000. *The Sociological Quarterly*, 44(3): 405–432.

Reid, S. (2022). Scotland powers ahead as homeowners rush to fit solar panels. *The Scotsman*, 16 February. Available at: https://www.scotsman. com/business/scotland-powers-ahead-as-homeowners-rush-to-fit-solar- panels-3570711 (Accessed 28/07/2022).

Renewables First. (2019). Netham Weir Hydroelectric Project: Supporting statement for a full planning application incorporating design and access statement, landscape and visual, heritage and noise assessments. Available at: http://bristolenergy.coop/wp-content/uploads/2020/03/18_05891_ f-planning_statement-2063497.pdf (Accessed 26/07/2022).

Reuters. (2021). Nuclear, coal, oil jobs pay more than those in wind, solar: report. *Reuters*, 6 April. Available at: https://www.reuters.com/ article/us-usa-energy-jobs-idUSKBN2BT2OT (Accessed 28/07/2022).

RIBA [Royal Institute of British Architects]. (2022). *Homes for Heroes: Solving the Energy Efficiency Crisis in England's Interwar Suburbs*. London: Royal Institute of British Architects.

Richter, B.D., Postel, S., Revenga, C., Scudder, T., Lehner, B., Churchill, A. and Chow, M. (2010). Lost in development's shadow: The downstream human consequences of dams. *Water Alternatives*, 3(2): 14–42.

Rignall, K.E. (2016). Solar power, state power, and the politics of energy transition in pre-Saharan Morocco. *Environment and Planning A: Economy and Space*, 48(3): 540–557.

Riofrancos, T. (2019). What green costs. *Logic*, 7 December. Available at: https://logicmag.io/nature/what-green-costs/ (Accessed 28/07/2022).

RMT [Rail, Maritime and Transport Workers, National Union of]. (2021). RMT uncovers illegal employment practices in the offshore supply chain. *RMT*, 13 May. Available at: https://www.rmt.org.uk/news/rmt-uncov ers-illegal-employment-practices-in-the-offshore/ (Accessed 28/07/2022).

Robinson, C. (2019). Energy poverty and gender in England: A spatial perspective. *Geoforum*, 104: 222–233.

Robinson, C., Bouzarovski, S. and Lindley, S. (2018). 'Getting the measure of fuel poverty': The geography of fuel poverty indicators in England. *Energy Research & Social Science*, 26: 79–93.

Robinson, C. and Mattioli, G. (2020). Double energy vulnerability: Spatial intersections of domestic and transport energy poverty in England. *Energy Research & Social Science*, 70: 101699.

Robinson, C. and Simcock, N. (2022). How can policy protect fuel poor households from rising energy prices? University of Liverpool: Heseltine Institute Policy Briefing, 2(16).

Roddis, P., Carver, S., Dallimer, M., Norman, P. and Ziv, G. (2018). The role of community acceptance in planning outcomes for onshore wind and solar farms: An energy justice analysis. *Applied Energy*, 226: 353–364.

Roddis, P., Roelich, K., Tran, K., Carver, S., Dallimer, M. and Ziv, G. (2020). What shapes community acceptance of large scale solar farms? A case study of the UK's first 'nationally significant' solar farm. *Solar Energy*, 209: 235–244.

Rodger, J. (2022). Woman so scared of rising energy bills that she's riding bus all day to stay warm. *Birmingham Live*, 19 March. Available at: https://www.birminghammail.co.uk/news/midlands-news/woman-scared-rising-energy-bills-23443295 (Accessed 28/07/2022).

Rodrik, D. (2022). An industrial policy for good jobs. Washington, DC.: The Brookings Institute.

Rose, M. and Felix, B. (2019). French strikers angry about pension reform cut power to homes, companies. *Reuters*, 18 December. Available at: https://www.reuters.com/article/us-france-protests-pensions-idUSKB N1YM0H6 (Accessed 09/12/2022).

Routledge, P., Chumbers, A. and Derickson, K.D. (2018). States of just transition: Realising climate justice through and against the state. *Geoforum*, 88: 78–86.

Royal British Legion. (2017). *Literature Review: UK Veterans and Homelessness*. London: Royal British Legion.

Rudolph, D., Haggett, C. and Aitken, M. (2018). Community benefits from offshore renewables: The relationship between different understandings of impact, community, and benefit. Environment and Planning C: Politics and Space, 36(1): 92–117.

Russell, A. and Firestone, J. (2021). What's love got to do with it? Understanding local genitive and affective responses to wind power projects. *Energy Research & Social Science*, 71: 101833.

Russell. S. (2022). Joanna Lumley raises concerns over offshore wind farm plans in joint letter. *Evening Standard*, 28 February. Available at: https://www.standard.co.uk/news/uk/joanna-lumley-ralph-fiennes-government-actor-suffolk-b985011.html (Accessed 28/07/2022).

Rutherford, N. (2021). Wind farm firms admit safety breaches over worker who froze to death. *BBC News*, 25 August. Available at: https://www.bbc.co.uk/news/uk-scotland-58331665 (Accessed 28/07/2022).

Rydin, Y., Lee, M. and Lock, S.J. (2015). Public engagement in decision-making on major wind energy projects. *Journal of Environmental Law*, 27(1): 139–150.

Rydin, Y., Natarajan, L., Lee, M. and Lock, S. (2018). Do local economic interests matter when regulating nationally significant infrastructure? The case of renewable energy infrastructure projects. *Local Economy*, 33(3): 269–286.

Ryser, S. (2019). The anti-politics machine of green energy development: The Moroccan Solar Project in Ouarzazate and its impact on gender ed local communities. *Land*, 8: 10.

Salisbury, J. (2022). This Morning criticised as 'dystopian' for offering viewers the chance to have their energy bills paid. *Evening Standard*, 5 September. Available at: https://www.standard.co.uk/news/uk/this-morning-energy-bills-itv-phillip-schofield-holly-willoughby-b1023250.html (Accessed 09/12/2022).

Samuel, R.I., Bloomfield, B. and Boanas, G. (1986). *The Enemy Within: Pit Villages and the Miners' Strike of 1984–1985*. Routledge and Kegan Paul: London.

Sanchez-Lopez, M.D. (2021). Territory and lithium extraction: The Great Land of Lipez and the Uyuni Salt Flat in Bolivia. *Political Geography*, 90: 102456.

Sapir, A., Schraepen, T. and Tagliapietra, S. (2022). Green public procurement: A neglected tool in the European Green Deal toolbox. *Interconomics: Review of European Economic Policy*, 57(3).

Sareen, S. and Haarstad, H. (2021). Decision-making and scalar biases in solar photovoltaics roll-out. *Current Opinion in Environmental Sustainability*, 51: 24–29.

Scheer, H. (2001). *A Solar Manifesto*. London: James & James.

Scheer, H. (2007). *Energy Autonomy: The Economic, Social and Technological Case for Renewable Energy*. London: Earthscan.

Scheiber, N. (2021). Building solar farms may not build the middle class. *The New York Times*, 16 July. Available at: https://www.nytimes.com/2021/07/16/business/economy/green-energy-jobs-economy.html (Accessed 28/07/2022).

Schlosberg, D. and Carruthers, D. (2010). Indigenous struggles, environmental justice, and community capabilities. Global Environmental Politics, 10(4): 12–35.

Schoetz, D. (2007). Wind farm? Not off my back porch. *ABC News*, 30 March. Available at: https://abcnews.go.com/US/story?id=2995334&page=1

Schools Energy Co-op (nd). Our schools. *Schools Energy Cooperative*. Available at: https://schools-energy-coop.co.uk/projects/ (Accessed 28/07/2022).

Schwanen, T. (2020). Low-carbon mobility in London: A just transition? *One Earth*, 2(2): 132–134.

Scott, D. and Smith, A.A. (2017). Sacrifice zones in the green energy economy: The new climate refugees. *Transnational Law and Contemporary Problems*, 26(2): 371–382.

Scudder, T. (2005). *The Future of Large Dams: Dealing with Social, Environmental, Institutional and Political Costs*. London: Routledge.

Seagreen. (2021). 87 per cent of Seagreen blades to be produced in UK. *Seagreen*, 28 June. Available at: https://www.seagreenwindenergy.com/post/87-of-seagreen-blades-to-be-produced-in-uk (Accessed 28/07/2022).

Seelye, K.Q. (2017). After 16 years, hopes for Cape Cod wind farm float away. *New York Times*, 19 December. Available at: https://www.nytimes.com/2017/12/19/us/offshore-cape-wind-farm.html (Accessed 28/07/2022).

Semuels, A. (2017). Do regulations really kill jobs? *The Atlantic*, 19 January. Available at: https://www.theatlantic.com/business/archive/2017/01/regulations-jobs/513563/ (Accessed: 25/07/2022).

Sen, A. (1999). Development as Freedom. Oxford: Oxford University Press.

Sheldon, P., Junankar, R. and De Rosa Pontello, A. (2018). *The Ruhr or Appalachia? Deciding the future of Australia's coal power workers and communities*. IRRC Report for CFMMEU Mining and Energy. Available at: https://www.ituc-csi.org/IMG/pdf/ruhrorappalachia_report_final.pdf (Accessed 27/07/2022).

Shell. (2022). First quarter 2022 results – May 5 2022. *Shell*, 5 May. Available at: https://www.shell.com/investors/results-and-reporting/quarterly-results/2022/q1-2022.html (Accessed 28/07/2022).

Sheppard, D. and Wilson, T. (2021). Fire at UK–France electricity subsea cable triggers new price surge. *Financial Times*, 15 December. Available at: https://www.ft.com/content/b5a5e29a-9556-4d07-882b-10b1749ef ced (Accessed 18/03/2023).

Sheppard, K. (2010). Cape Wind site not so sacred after all? *Mother Jones*, 26 February. Available at: https://www.motherjones.com/politics/2010/02/cape-wind-site-not-so-sacred-after-all/ (Accessed 28/07/2022).

Shetland News. (2020). Agreement reached on £2.2m-a-year Viking Energy community benefit fund. *Shetland News*, 29 July. Available at: https://www.shetnews.co.uk/2020/07/29/agreement-reached-on-2-2m-a-year-viking-energy-community-benefit-fund/ (Accessed 28/07/2022).

Shildrick, T. (2018). Lessons from Grenfell: Poverty propaganda, stigma and class power. *Sociological Review*, 66(4): 783–798.

Shrubsole, G. (2020). *Who Owns England?: How We Lost Our Land and How to Take it Back*. Glasgow: William Collins.

Siemens Gamesa. (2022). Press release – Revolutionary RecyclableBlades: Siemens Gamesa technology goes full-circle at RWE's Kaskasi offshore wind power project. *Siemens Gamesa*, 1 August. Available at: https://www.siemensgamesa.com/en-int/newsroom/2022/07/080122-siemens-gam esa-press-release-recycle-wind-blade-offshore-kaskasi-germany (Accessed 09/12/2022).

Sigrin, B., Sekar, A. and Tome, E. (2022). The solar influence next door: Predicting low-income solar referrals and leads. *Energy Research & Social Science*, 86: 102417.

Sillars, J. (2022). Britishvolt workers take 'substantial' pay cut as firm seeks 'more secure funding position'. *Sky News*, 2 November. Available at: https://news.sky.com/story/britishvolt-workers-agree-pay-cut-as-firm-seeks-more-secure-funding-position-12736145 (Accessed 09/12/2022).

Simcock, N. (2016). Procedural justice and the implementation of community wind energy projects: A case study from South Yorkshire, UK. *Land Use Policy*, 59: 467–477.

Simcock, N., Frankowski, J. and Bouzarovski, S. (2021). Rendered invisible: Institutional misrecognition and the reproduction of energy poverty. Geoforum, 124: 1–9.

Singh, J.N. (2021). Mining our way out of the climate change conundrum? The power of a social justice perspective. *Wilson Center*, October 2021.

SJTC [Scottish Just Transition Commission]. (2020). Just Transition Commission: Interim Report. *Scottish Government*, 27 February. Available at: https://www.gov.scot/publications/transition-commission-interim-rep ort/ (Accessed 16/03/2023).

SJTC. (2021). Just Transition Commission: A National Mission for a fairer, greener Scotland. *Scottish Government*, 23 March. Available at: https://www.gov.scot/publications/transition-commission-national-mission-fairer-greener-scotland/pages/3/ (Accessed 25/07/2022).

SJTC. (2022). Making the Future – second Just Transition Commission: initial report. *Scottish Government*, 14 July. Available at: https://www.gov.scot/publications/making-future-initial-report-2nd-transition-commission/pages/2/ (Accessed 09/12/2022).

SLRRG [Scottish Land Reform Review Group]. (2014). *The Land of Scotland and the Common Good: Report of the Land Reform Review Group*. Presented to Scottish Ministers in 2014. Scottish Government, 23 May. Available at: https://www.gov.scot/publications/land-reform-review-group-final-report-land-scotland-common-good/pages/61/ (Accessed 26/07/2022).

Smith, A. (2014). *Socially Useful Production*. STEPS Working Paper 58. Brighton: STEPS Centre.

Smith, N. (2021). The left's NIMBY war against renewable energy. *Bloomberg UK*, 12 September. Available at: https://www.bloomberg.com/opinion/articles/2021–09–12/the-left-s-nimby-war-against-renewable-energy#xj4y7vzkg (Accessed 28/07/2022).

Smith, O. (2022). Prepay energy customers disconnect over price rises. *BBC News*, 29 April. Available at: https://www.bbc.co.uk/news/business-61270970.amp (Accessed 28/07/2022).

Smitherman, E. (2022). Group fighting proposed Rampion 2 wind farm off West Sussex coast launch petition. *The Argus*, 16 January. Available at: https://www.theargus.co.uk/news/19848605.group-fighting-proposed-wind-farm-off-west-sussex-coast-launch-petition/ (Accessed 26/07/2022).

Snell, C., Bevan, M. and Thomson, H. (2015). Justice, fuel poverty and disabled people in England. *Energy Research & Social Science*, 10: 123–132.

Snell, C.J., Bevan, M.A. and Gillard, R. (2018). Policy pathways to justice in energy efficiency. Research Report, *UK Energy Research Centre*. Available at: https://ukerc.ac.uk/publications/policy-pathways-to-justice-energy-efficiency/ (Accessed 27/07/2022).

Snell, D. (2018). 'Just transition'? Conceptual challenges meet stark reality in a 'transitioning' coal region in Australia. *Globalizations*, 15(4): 550–564.

Snell, D. and Fairbrother, P. (2010). Unions as environmental actors. *Transfer: European Review of Labour and Research*, 16(4): 411–424.

Snyder, B.F. (2018). Vulnerability to decarbonization in hydrocarbon-intensive counties in the United States: A just transition to avoid post-industrial decay. Energy Research & Social Science, 42: 34–43.

Social Metrics Commission. (2020). *Measuring Poverty 2020: A report of the Social Metrics Commission*. London: Social Metrics Commission.

Solar United Neighbours, Institute for Local Self-Reliance, and Initiative for Energy Justice. (2020). *30 Million Solar Homes: A Vision for an Equitable Economic Recovery. Built on Climate Protection and Energy Democracy.* Washington, DC: Solar United Neighbors.

Sommerfeld, J., Buys, L. and Vine, D. (2017). Residential consumers' experiences in the adoption and use of solar PV. *Energy Policy*, 105: 10–16.

Sonnberger, M. and Ruddat, M. (2017). Local and socio-political acceptance of wind farms in Germany. *Technology in Society*, 51: 56–65.

Sørensen, H.C., Hansen, L.K. and Larsen, J.H.M. (2002). Middelgrunden 40 MW offshore wind farm Denmark-lessons learned. *Renewable realities–offshore wind technologies*. Orkney, UK; 2002. Available at: https://citeseerx.ist.psu.edu/viewdoc/download?doi=10.1.1.567.9273&rep=rep1&type=pdf (Accessed 26/07/2022).

Soukhaphon, A., Baird, I.G., Hogan, Z.S. (2021). The impacts of hydropower dams in the Mekong River basin: A review. *Water*, 13: 265.

Sovacool, B.K. (2015). Fuel poverty, affordability and energy justice in England: Policy insights from the Warm Front programme. *Energy*, 93: 361–371.

Sovacool, B.K. (2021). When subterranean slavery supports sustainability transitions? power, patriarchy, and child labor in artisanal Congolese cobalt mining. *The Extractive Industries and Society*, 8: 271–293.

Sovacool, B.K. and Dworkin, M.H. (2015). Energy justice: Conceptual insights and practical applications. *Applied Energy*, 142: 435–444.

Sovacool, B.K., Baker, L., Martiskainen, M. and Hook, A. (2019a). Processes of elite power and low-carbon pathways: Experimentation, financialisation and dispossession. *Global Environmental Change*, 59: 101985.

Sovacool, B.K., Martiskainen, M., Hook, A. and Baker, L. (2019b). Decarbonization and its discontents: A critical energy perspective on four low-carbon transitions. Climatic Change, 155: 581–619.

Sovacool, B.K., Hook, A., Martiskainen, M., Brock, A. and Turnheim, B. (2020). The decarbonisation divide: Contextualizing landscapes of low-carbon exploitation and toxicity in Africa. *Global Environmental Change*, 60: 102028.

Sovacool, B.K., Turnheim, B., Hook, A., Brock, A. and Martiskainen, M. (2021). Dispossessed by decarbonisation: Reducing vulnerability, injustice, and inequality in the lived experience of low-carbon pathways. *World Development*, 137: 105116.

Sovacool, B.K., Lacey Barnacle, M., Smith, A. and Brisbois, M.C. (2022). Towards improved solar energy justice: Exploring the complex inequities of houserhold adoption of photovoltaic panels. *Energy Policy*, 164: 112868.

Sperling, K. (2017). How does a pioneer community energy project succeed in practice? The case of the Samsø Renewable Energy Island. *Renewable and Sustainable Energy Reviews*, 81: 884–897.

Spinazzè, A., Cattaneo, A., Monticelli, D., Recchia, S., Rovelli, S., Fustinoni, S. and Cavallo, D.M. (2015). Occupational exposure to arsenic and cadmium in thin-film solar cell production. *Ann Occup Hyg*, 59(5): 572–585.

Spivey, H. (2020) Governing the fix: Energy regimes, accumulation dynamics, and land use changes in Japan's solar photovoltaic boom. *Annals of the American Association of Geographers*, 110(6): 1690–1708.

Stedman, R. (2002). Toward a social psychology of place: predicting behaviour from place-based cognitions, attitude, and identity. *Environment and Behaviour*, 34: 561–581

Stephens, J.C. (2021). *Diversifying Power: Why we Need Antiracist, Feminist Leadership on Climate and Energy*. Washington DC.: Island Press.

Stevis, D. (2018a) (Re)claiming just transition. Just Transition Research Collaborative (JTRC). Available at: https://medium.com/just-transitions/stevis-e147a9ec189a (Accessed 25/07/2022).

Stevis, D. (2018b). US labour unions and green transitions: Depth, breadth and worker agency. *Globalizations*, 15(4): 454–469.

Stevis, D. (2019). Labor union and green transitions in the US: Contestations and explanations. ACW Working Paper, 108.

Stevis, D. and Felli, R. (2015). Global labour unions and just transition to a green economy. *International Environmental Agreements*, 15: 29–43.

Stevis, D., Uzzell, D. and Räthzel, N. (2018). The labour–nature relationship: Varieties of labour environmentalism. *Globalizations*, 15(4): 439–453.

Stewart, F. (2022a). Power for the people. *Medium*, 18 January. Available at: https://fraserjfstewart-17.medium.com/energy-in-the-uk-21f6ac1fe a90 (Accessed 28/07/2022).

Stewart, F. (2022b). Friends with benefits: How income and peer diffusion combine to create an inequality 'trap' in the uptake of low-carbon technologies. *Energy Policy*, 163: 112832.

Stokes, L. (2020). *Short Circuiting Policy: Interest Groups and the Battle over Clean Energy and Climate Policy in the American States*. Oxford: Oxford University Press.

Strietska-Ilina, O., Hoffman, C., Durán Haro, M. and Jeon, S. (2011). *Skills for Green Jobs: A Global View*. Geneva: International Labour Organization.

STUC [Scottish Trade Union Congress] (2019). *Broken Promises and Offshored Jobs: STUC report on employment in the lowcarbon and renewable energy economy*. Glasgow: Scottish Trade Unions Congress.

STUC and Transition Economics. (2021). *Green Jobs in Scotland*. Glasgow: Scottish Trade Unions Congress.

Sturgis, P. and Allum, N. (2004). Science in society: Re-evaluating the deficit model of public attitudes. *Public Understanding of Science*, 13(1): 55–74.

Sugar, K.H. (2021). *Governing Low-carbon and Inclusive Transitions in the City: A Case Study of Nottingham, UK*. PhD thesis, University of Glasgow.

Sugar, K. and Webb, J. (2022). Value for money: Local authority action on clean energy for net zero. *Energies*, 15(12): 4359.

Sulich, A., Rutkowska, M. and Poplawski, L. (2020). Green jobs, definitional issues, and the employment of young people: An analysis of three European Union countries. *Journal of Environmental Management*, 6262: 110314.

Sultana, F. (2022). The unbearable heaviness of climate coloniality. *Political Geography*, 102638.

Sunter, D.A., Castellanos, S. and Kammen, D.M. (2019). Disparities in rooftop photovoltaics deployment in the United States by race and ethnicity. *Nature Sustainability*, 2: 71–76.

Sussex Wildlife Trust. (2021). Rampion 2 Consultation. Available at: https://sussexwildlifetrust.org.uk/campaign/rampion-2 (Accessed 26/07/2022).

Sutter, C. and Parreño, J.C. (2007). Does the current Clean Development Mechanism (CDM) deliver its sustainable development claim? An analysis of officially registered CDM projects. *Climatic Change*, 84: 75–90.

Sutton, I. (2021). Germany: will the end of feed-in tariffs mean the end of citizens-as-energy-producers. *Energypost.eu*, 3 June. Available at: https://energypost.eu/germany-will-the-end-of-feed-in-tariffs-mean-the-end-of-citizens-as-energy-producers/ (Accessed 28/07/2022).

Swinford, S. (2022). Rishi Sunak cuts energy bills by £400 to tackle cost of living crisis: Oil and gas firms hit with 25 per cent windfall tax. *The Times*, 26 May. Available at: https://www.thetimes.co.uk/article/rishi-sunak-to-cut-energy-bills-by-hundreds-of-pounds-with-windfall-tax-on-oil-and-gas-firms-p9ntgs6bp (Accessed 28/07/2022).

Szulecki, K. and Overland, I. (2020). Energy democracy as a process, an outcome and a goal: A conceptual review. *Energy Research and Social Science*, 69: 101768.

tado. (2020). UK homes losing heat up to three times faster than European neighbours. *Tado*, 20 February. Available at: https://www.tado.com/gb-en/press/uk-homes-losing-heat-up-to-three-times-faster-than-european-neighbours (Accessed 28/07/2022).

Taher, A. and Hind, K. (2022). Amanda Holden backs war on offshore wind farm with 1,066ft turbines taller than the Eiffel Tower which would blight West Sussex coast. *MailOnline*, 15 January. Available at: https://www.dailymail.co.uk/news/article-10406375/Amanda-Holden-objects-shore-wind-farm-near-Sussex-home.html (Accessed 15/12/2022).

Takesada, N. (2009). Japanese experience of involuntary resettlement: long-term consequences of resettlement for the construction of the Ikawa Dam. *Int. J. Water Resourc. Dev.*, 25(3): 419–430.

Tapsfield, J. and Heffer,G. (2022). Second home owners will save £400 on energy bills for EVERY property under Rishi's bailout – with Chancellor himself set to benefit (but he's donating cash to charity). *MailOnline*, 26 May. Available at: https://www.dailymail.co.uk/news/article-10857985/Second-home-owners-save-400-energy-bills-property-Rishi-Sunaks-bail out.html (Accessed 28/07/2022).

Teron, L. and Ekoh, SS. (2018). Energy democracy and the city: Evaluating the practice and potential of municipal sustainability planning. *Front. Commun.* 3(8).

The Spectator. (2021). How the Tories have fuelled Britain's energy crisis. *The Spectator*, 25 September. Available at: https://www.spectator.co.uk/article/how-the-tories-have-fuelled-britains-energy-crisis/ (Accessed 18/03/2023).

The White House (USA). (2021). *Building Resilient Supply Chains: Revitalizing American Manufacturing and Fostering Broad-Based Growth: 100-Day Reviews under Executive Order 14017*. Washington, DC: The White House.

Thomas, A. and Doerflinger, N. (2020). Trade union strategies on climate change mitigations: Between opposition, hedging and support. *European Journal of Industrial Relations*, 26(4): 383–399.

Thomas, D. (2022). Energy crisis pushing people onto prepayment meters, says Uswitch. *BBC News*, 9 December. Available at: https://www.bbc.co.uk/news/business-63378460 (Accessed 09/12/2022).

Thompson, M. (2020). What's so new about New Municipalism? *Progress in Human Geography*, 45(2): 317–342.

Thrive Renewables. (no date.). Community benefit programme. Thrive Renewables. Available at: https://www.thriverenewables.co.uk/our-imp act/community-benefit-programme/ (Accessed 28/07/2022).

Tolnov Clausen, L. and Rudolph, D.P. (2019). (Dis)embedding the wind: On people-climate reconciliation in Danish wind power planning. *Journal of Transdisciplinary Environmental Studies*, 17(1): 5–21.

Tomer, A., Kane, J.W. and George, C. (2021). How renewable energy jobs can uplift fossil fuel communities and remake climate politics. *Brookings Institute*, 23 February. Available at: https://www.brookings.edu/research/how-renewable-energy-jobs-can-uplift-fossil-fuel-communities-and-rem ake-climate-politics/ (Accessed 27/07/2022).

Trabish, H.K. (2014). Cape Wind opponents heading back to appeals court – again. *Utility Dive*, 2 September. Available at: https://www.utilitydive.com/news/cape-wind-opponents-heading-back-to-appeals-court-again/303762/ (Accessed 28/07/2022).

Treacy, M. (2010). Reporting Vestas on Isle of Wight became 'way of life'. *BBC News*, 21 July. Available at: http://news.bbc.co.uk/local/hampsh ire/hi/people_and_places/newsid_8811000/8811650.stm (Accessed 18/03/2023).

Tuan Y-F. (1977). *Space and Place: The Perspective of Experience.* Minneapolis: University of Minnesota Press.

Turner, C. (2022). NHS pays for patients' energy bills in winter pilot to cut hospital admissions. *The Telegraph,* 22 November. Available at: https://www.telegraph.co.uk/news/2022/11/22/nhs-prescribes-warmth-respiratory-patients-pays-energy-bills (Accessed 13/12/2022).

UK Committee on Climate Change. (2021). 2021 Progress Report to Parliament. London: UK Climate Change Committee.

UK Committee on Climate Change. (2022). 2022 Progress Report to Parliament. London: UK Climate Change Committee.

UK Green Jobs Taskforce. (2021). Green Jobs Taskforce: Report to Government, Industry and the Skills sector. London: UK Government.

UK National Infrastructure Planning. (2021). Cleve Hill Solar Park. London: UK Government. Available at: https://infrastructure.planninginspectorate.gov.uk/projects/south-east/cleve-hill-solar-park/?ipcsection=overview (Accessed 16/03/2023).

UK Parliament. (2019). Local government finance and the 2019 Spending Review: Local government funding. Available at: https://publications.parliament.uk/pa/cm201719/cmselect/cmcomloc/2036/203605.htm (Accessed 28/07/2022).

UK Warehousing Association. (2022). Time to act on extortionate and obstructive electricity grid if we are to tackle the energy crisis, say UK warehouse owners. *UK Warehousing Association,* 7 September. Available at: https://www.ukwa.org.uk/market-intel/time-to-act-on-extortionate-and-obstructive-electricity-grid-if-we-are-to-tackle-the-energy-crisis-say-uk-warehouse-owners/ (Accessed 09/12/2022).

UN [United Nations]. (2021). *Theme Report on Energy Transition: Towards the Achievement of SDG 7 and Net-Zero Emissions.* New York: United Nations.

UNFCCC [United Nations Framework Convention on Climate Change]. (2016). 'Paris Agreement'. In UNFCCC COP Report Number 21, Addendum, at 21, U.N. Do. FCCC/CP/2015/10/Add.1. Available at: https://unfccc.int/resource/docs/2015/cop21/eng/10a01.pdf

UNFCCC. (2020). *Just Transition of the Workforce, and the Creation of Decent Work and Quality Jobs.* UNFCCC Technical Paper. Available at: https://unfccc.int/documents/226460 (Accessed 09/12//2022).

Unite the Union. (2017). Workers stage work-in at BiFab yards. *Unite the Union,* 13 November. Available at: https://www.unitetheunion.org/what-we-do/unite-in-your-region/scotland/latest-news/workers-stage-work-in-at-bifab-yards/ (Accessed 27/07/2022).

US Department of Energy. (2022). *U.S. Energy & Employment Jobs Report (USEER).* Washington, D.C.: United States Office of Policy.

USGS [United States Geological Survey]. (2022). Mineral Commodity Summaries 2022. Washington, DC: US Department of the Interior.

Vachon, T. (2019). *The green transition: Renewable energy technology, climate change mitigation and the future of work in New Jersey*. Prepared for the New Jersey Governor's Task Force on the Future of Work. Available at: https://fowtf.innovation.nj.gov/downloads/resources/Vachon_Climate_Jobs_Report_RCS.pdf (Accessed 27/07/2022).

Vachon, T.E. and Brecher, J. (2016). Are union members more or less likely to be environmentalist? Some evidence from two national surveys. *Labor Studies Journal*, 41(2): 185–203.

Van Bommel, N. and Höffken, J.I. (2021). Energy justice within, between and beyond European community energy initiatives: A review. *Energy Research & Social Science*, 79: 102157.

Van de Ven, D-J., Capellan-Peréz, I., Arto, I., Cazcarro, I., de Castro, C., Patel, P. and Gonzalez-Eguino, M. (2021). The potential land requirements and related land use change emissions of solar energy. *Scientific Reports*, 11: 2907.

Van der Horst, D. and J. Evans. (2010). Carbon claims and energy landscapes: exploring the political ecology of biomass. *Landscape Research*, 35(2): 173–193.

Van Veelen, B. (2018). Negotiating energy democracy in practice: Governance processes in community energy projects. *Environmental Politics*, 27(4): 644–665.

Van Veelen, B. and Van der Horst, D. (2018). What is energy democracy? Connecting social science energy research and political theory. *Energy Research & Social Science*, 46: 19–28.

Van Wijk, J., Fischhendler, I., Rosen, G. and Herman, L. (2021). Penny wise or pound foolish? Compensation schemes and the attainment of community acceptance in renewable energy. *Energy Research & Social Science*, 81: 102260.

Vaughan, R. and Waugh, P. (2022). Boris Johnson plans vast nuclear energy expansion but waters down onshore wind ambitions. *iNews*, 6 April. Available at: https://inews.co.uk/news/politics/boris-johnson-nuclear-energy-expansion-onshore-wind-ambitions-1561194 (Accessed 28/07/2022).

Vidal, J. (2012). Wind turbines bring in 'risk-free' millions for rich landowners. *The Guardian*, 28 February. Available at: https://www.theguardian.com/environment/2012/feb/28/wind farms -risk-free-millions-for-landowners (Accessed 28/07/2022).

Voskoboynik, D.M. and Andreucci, D. (2021). Greening extractivism: Environmental discourses and resource governance in the 'Lithium Triangle'. *Environment and Planning E: Nature and Space*, 5(2): 787–809.

Vyn, R. (2019). Building wind turbines where they're not wanted brings down property values. *The Conversation*, 7 January. Available at: https://theconversation.com/building-wind-turbines-where-theyre-not-wanted-brings-down-property-values-106690 (Accessed 26/07/2022).

Wajsbrot, S. (2022). EDF: l'Etat confirme la nationalisation à 100 per cent. *Les Echos*, 6 July. Available at: https://www.lesechos.fr/industrie-services/energie-environnement/letat-confirme-la-nationalisation-dedf-a-100-1775016 (Accessed 28/07/2022).

Walker, C. and Baxter, J. (2017a). Procedural justice in Canadian wind energy development: a comparison of community-based and technocratic siting processes. *Energy Research & Social Science*, 29: 160–169.

Walker, C. and Baxter, J. (2017b). "It's easy to throw rocks at a corporation": wind energy development and distributive justice in Canada. *J. Environ. Policy Plan.* (2017): 1–15.

Walker, G. (2009). Beyond distribution and proximity: Exploring the multiple spatialities of environmental justice. Antipode, 41(4): 614–636.

Walker, G. and Devine-Wright, P. (2008). Community renewable energy: What should it mean? *Energy Policy*, 36(2): 497–500.

Walker, G., Devine-Wright, P., Hunter, S., High, H. and Evans, B. (2010). Trust and community: Exploring the meanings, contexts and dynamics of community renewable energy policy. Energy Policy, 38: 2655–2663.

Walker, G., Devine-Wright, P., Barnett, J., Burningham, K., Cass, N., Devine-Wright, H., Speller, G., Barton, J., Evams. N., heath, Y., Infield, D., Parks, J. and Theobald, K. (2011). 'Symmetries, expectations, dynamics and contexts: A framework for understanding public engagement with renewable energy projects'. In: Devine-Wright, P. (ed.). *Renewable Energy and the Public: From NIMBY to Participation*. London: Earth Scan from Routledge.

Walker, G. and Day, R. (2012). Fuel poverty as injustice: Integrating distribution, recognition and procedure in the struggle for affordable warmth. *Energy Policy*, 49: 69–75.

Walker, L., Fernandes, F. and Roberts, G. (2014). The Planning Act 2008, Rampion Offshore Wind Farm and connection works: Examining Authority's Report of Findings and Conclusions and Recommendation to the Secretary of State for Energy and Climate Change. *The Planning Inspectorate, UK*. Available at: https://infrastructure.planninginspectorate.gov.uk/wp-content/ipc/uploads/projects/EN010032/EN010032-001704-Rampion per cent20Recommendation per cent20Report.pdf (Accessed 26/07/2022).

Walne, T. (2022). Vastly expensive roll-out of smart energy meters described as a 'waste of money' – because equipment will become obsolete. *This is Money*, 15 January. Available at: https://www.thisismoney.co.uk/money/bills/article-10405685/ALL-smart-meters-need-replaced.html (Accessed 28/07/2022).

Warbroek, B. and Hoppe, T. (2017). Modes of governing and policy of local and regional governments supporting local low-carbon energy initiatives; Exploring the cases of the Dutch regions of Overijssel and Fryslân. *Sustainability*, 9(1): 75.

Ward, P. (2017). Windfarm worker plunges to his death from turbine near Glasgow just weeks after tragedy at Ayrshire facility. *Daily Record*, 31 March. Available at: https://www.dailyrecord.co.uk/news/scottish-news/windf arm-worker-plunges-death-turbine-10134003 (Accessed 28/07/2022).

Warren, C.R. and McFadyen, M. (2010). Does community ownership affect public attitudes to wind energy? A case study from south-west Scotland. *Land Use Policy*, 27: 204–213.

Watt, H. (2018). Crofters on Lewis fight EDF and Wood Group's windfarm proposal. *The Guardian*, 4 February. Available at: https://www.theguard ian.com/uk-news/2018/feb/04/windfarm-crofters-lewis-fight-edf-wood-group-scottish (Accessed 28/07/2022).

Watts, L. (2018). *Energy at the End of the World: An Orkney Islands Saga.* Cambridge, MA.: The MIT Press.

Way, R., Ives, M., Mealy, P. and Doyne Farmer, J. (2021). Empirically grounded technology forecasts and the energy transition. Instituyte of New Economic Thinking, Oxford Martin School: INET Oxford Working Paper No. 2021–01.

Wearmouth, R. (2022). Labour must "face reality" on the energy crisis and back fracking, says GMB boss. *The New Statesman*, 22 September. Available at: https://www.newstatesman.com/politics/labour/2022/09/labour-frack ing-energy-crisis-gmb-union (Accessed 09/12/2022).

Wearn, R. and Masud, F. (2022). £400 energy support vouchers going unclaimed. *BBC News*, 31 October. Available at: https://www.bbc.co.uk/news/business-63412380.

Weaver, M. and Morris, S. (2009). Staff occupy Isle of Wight wind turbine plant in protest against closure. *The Guardian*, 21 July. Available at: https://www.theguardian.com/environment/2009/jul/21/wind-turbine-factory-occupation (Accessed 28/07/2022).

Weber, G. and Cabras, I. (2017). The transition of Germany's energy production, green economy, low-carbon economy, socio-environmental conflicts, and equitable society. *Journal of Cleaner Production*, 167: 1222–1231.

Webster, B. (2022). New homes built with gas boilers after developers lobby against green rules. *openDemocracy*, 3 October. Available at: https://www. opendemocracy.net/en/new-build-homes-gas-boilers-heat-pumps-dev elopers-lobby-government/ (Accessed 18/03/2023).

Webster, R. and Shaw, C. (2020). *Broadening Engagement with Just Transition.* Oxford: Climate Outreach.

Weghmann, V. and Hall, D. (2021). A publicly owned energy industry could help tackle energy poverty and increase renewables. *The Conversation*, 12 October. Available at: https://theconversation.com/a-publicly-owned-energy-industry-could-help-tackle-energy-poverty-and-increase-renewables-169186 (Accessed 28/07/2022).

Weißermel, S. (2021). The (im)possibilitiy of agonistic politics: The Belo Monte dam and the symbolic order of dispossession. *Geoforum*, 126: 91–100.

Welsh, M. (2021). Cold homes, hot planet. *New Economics Foundation*, 19 August. Available at: https://neweconomics.org/2021/08/cold-homes-hot-planet (Accessed 28/07/2022).

We Own It. (no date.). Public ownership saves billions. weownit.org.uk. Available at: https://weownit.org.uk/why-public-ownership/nationalisation-saves-billions (Accessed 28/07/2022).

Western Sahara Resource Watch. (2021). *Greenwashing Occupation: How Morocco's Renewable Energy Projects in Occupied Western Sahara Prolong the Conflict the Last Colony in Africa*. Brussels: Western Sahara Resource Watch.

Whitaker, A. (2017). BiFab: Inside the Occupation. *The Herald*, 19 November. Available at: https://www.heraldscotland.com/news/15670 209.bifab-inside-occupation/.

Whitefield, G. (2021). Port of Blyth launches clean energy terminal to capitalise on booming renewables market. *Business Live*, 15 February. Available at: https://www.business-live.co.uk/ports-logistics/port-blyth-launches-clean-energy-19830114 (Accessed 28/07/2022).

Whyte, K. (2018). 'The recognition paradigm of environmental injustice'. In: Holifeld, R., Chakraborty, J. and Walker, G. (eds). *The Routledge Handbook of Environmental Justice*. Abingdon, Oxon: Routledge.

Wickham, A. and Gillespie, T. (2022). UK sees up to £170 billion excess profits for energy firms. *Bloomberg UK*, 30 August. Available at: https://www.bloomberg.com/news/articles/2022-08-30/uk-predicts-up-to-170-billion-excess-profits-for-energy-firms (Accessed 09/12/2022).

Wien Stadwerke. (2020). *Laying the groundwork for a secure future: Financial report 2020*. Vienna: Wien Stadwerke.

Willand, N. and Horne, R. (2018). 'They are grinding us into the ground': The lived experience of (in)energy justice amongst low-income older households. *Applied Energy*, 226: 61–70.

Willand, N., Middha, B. and Walker, G. (2021). Using the capability approach to evaluate energy vulnerability policies and initiatives in Victoria, Australia. *Local Environment*, 26(9): 11091127.

Williams, M. (2022). Revealed: E.ON tried to get Kwarteng to cut its taxes while bills soared. *openDemocracy*, 11 October. Available at: https://www.opendemocracy.net/en/dark-money-investigations/eon-lobbying-beis-kwasi-kwarteng-tax-cuts-regulation/ (Accessed 09/12/2022).

Wilson, K. (2018). Bristol City Council drops Bristol Energy as green energy provider – despite owning it. *Bristol Live*, 25 April. Available at: https://www.bristolpost.co.uk/news/bristol-news/bristol-city-council-drops-bristol-1499376 (Accessed 28/07/2022).

Windemer, R. (2022). The impact of the 2015 onshore wind policy change for local planning authorities in England: Preliminary survey results. Available at: https://uwe-repository.worktribe.com/output/9206381 (Accessed 26/07/2022).

Windsong, E.A. (2021). White and Latino differences in neighbourhood emotional connections and the racialization of space. *Sociological Focus*, 54(3): 167–185.

Winemiller, K.O., McIntyre, P.B., Castello, L., Fluet-Chouinard, E., Giarrizzo, T., Nam, S. et al (2016). Balancing hydropower and biodiversity in the Amazon, Congo, and Mekong. *Science*, 351(6269): 128–129.

Wolsink, M. (2007). Wind power implementation: The nature of public attitudes: Equity and fairness instead of 'backyard motives. *Renewable and Sustainable Energy Reviews*, 11: 118–1207.

Woods, B. (2021). Smart meters and GPS under threat as old mobile networks get switched off. *The Telegraph*, 8 December. Available at: https://www.telegraph.co.uk/business/2021/12/08/2g-3g-phased-2033-blow-smart-meter-roll/ (Accessed 28/07/2022).

World Bank. (2017). *The Growing Role of Minerals and Metals for a Low-carbon Future*. Washington, DC.: World Bank Working Paper, 117581.

World Commission on Dams. (2000). *Dams and Development: A New Framework for Decision-Making*. London: World Commission on Dams.

WRI [World Resources Institute]. (2021). Spain's national strategy to transition coal-dependent communities. *World Resources Institute*, 23 December. Available at: https://www.wri.org/update/spains-national-strategy-transition-coal-dependent-communities (Accessed 28/07/2022).

WWEA [World Wind Energy Association]. (2022). World mark for wind power saw another record year in 2021: 97,3 gigawatt of new capacity add. *World Wind Energy* Association, 178 March. Available at: https://wwindea.org/world-market-for-wind-power-saw-another-record-year-in-2021-973-gigawatt-of-new-capacity-added/ (Accessed 09/12/2022).

Wüstenhagen, R., Wolsink, M. and Bürer, M.J. (2007). Social acceptance of renewable energy innovation: An introduction to the concept. *Energy Policy*, 35(5): 2683–2691.

Xlinks. (2022). The Morocco–UK Power Project. *Xlinks*. Available at: https://xlinks.co/morocco-uk-power-project/ (Accessed 28/07/2022).

Xu, Y., Li, J., Tan, Q., Peters, L.P. and Yang, C. (2018). Global status of recycling waste solar panels: A review. *Waste Management*, 75: 450–458.

Yenneti, K., Day, R. and Golubchikov, O. (2016). Spatial justice and the land politics of renewables: Dispossessing vulnerable communities through solar energy mega projects. *Geoforum*, 76: 90–99.

Zeller, Jr, T. (2010). For those near, the miserable hum of clean energy. *The New York Times*, 5 October. Available at: https://www.nytimes.com/2010/10/06/business/energy-environment/06noise.html (Accessed 28/07/2022).

Ziv, G., Baran, E., Nam, S., Rodríguez-Iturbe, I. and Levin, S.A. (2012). Trading-off fish biodiversity, food security, and hydropower in the Mekong River Basin. *Proceedings of the National Academy of Sciences*, 109 (15) 5609–5614.

Zografos, C. and Robbins, P. (2020). Green sacrifice zones, or why a Green New Deal cannot ignore the costs shifts of just transitions. *One Earth*, 3(5): 543–546.

Index

References to illustrations appear in *italic* type.